Industrial Electronics for Technicians

J.A. Sam Wilson

and

Joseph Risse

An Imprint of
Howard W. Sams & Company
Indianapolis, Indiana

REVISED FIRST EDITION, 1994

PROMPT® Publications is an imprint of Howard W. Sams & Company, 2647 Waterfront Parkway, East Drive, Suite 300, Indianapolis, IN 46214-2041.

This book was originally developed and published by TAB Books, Blue Ridge Summit, PA 17294-0214. Portions of this book originally appeared in *Industrial Electronics CET Exam Study Guide* by J.A. Sam Wilson, also published by TAB Books.

International Standard Book Number: 0-7906-1058-2

Cover Design by: Phil Velikan

Printed in the United States of America

9 8 7 6 5 4 3

Contents

About the Authors

J. A. Sam Wilson has written numerous books covering all aspects of the electronics field, and his articles have appeared in *Electronic Servicing & Technology Magazine*. He has a Radio Amateur General License, a lifetime First Class Commercial FCC License with Radar Endorsement, and a Journeyman Certified Electronics Technicians rating in consumer electronics. Sam has served as the Director of Technical Publications for NESDA and as CET Test Consultant for ISCET. His academic credits include a diploma from Capitol Radio Engineering Institute, a bachelor's degree from Long Beach State College, and a master's degree from Kent State University, as well as three diplomas from correspondence schools. Currently, Sam is a full-time freelance technical writer and consultant.

Joseph Risse worked for several years as a chief engineer at radio/TV commercial broadcast stations. His career is now devoted to developing and writing courses, lessons, and laboratory experiments in both self-study and industrial electronics for International Correspondence Schools and other independent study schools. Joe has also published several books on electronic test equipment. He has a lifetime First Class Commercial FCC License with Radar Endorsement, and holds a B.A. in natural science/mathematics from A. Edison College. Joe also has degrees in mathematics, physics, and electronics from the University of Scranton, as well as diplomas from Marywood College and International Correspondence Schools.

Preface

Industrial electronics is a specialty. There are components and circuits and, of course, systems that are used extensively in this field. In this book, *components* are the parts that make a circuit. Resistors, capacitors, and transistors are examples of components.

A *circuit* is a combination of components that performs a specific job. Oscillators and amplifiers are examples of circuits. *Systems* are combinations of circuits. A numerical control system has many circuits.

Some examples do not exactly fit into any of these categories. For example, a power supply can be a very important circuit that is part of a computer. It may also be a complete system used on a technician's bench.

The subjects in this book are covered with the assumption that the reader has passed through courses in fundamentals such as Ohm's law and general electronic circuitry. In some cases those basic subjects will be reviewed in the book as an introduction to a more advanced discussion.

In order to take the Journeyman CET Test you must first pass the Associate Level Test. That test is on general and basic subjects that are prerequisite for understanding this book. Also, it is necessary to have four years of working experience in the field in order to take a Journeyman Test for any subject.

Your work experience in industrial electronics may have been in some specialty, so, it is a good idea to review a broader range of electronic concepts before taking the test. This is one of the purposes of this book.

The book will also be helpful to anyone with electronic experience or training who is preparing to go into the industrial electronics field. Some specialties in industrial electronics are very broad—in fact, too broad to be covered in a book like this. For example, computers are now being used extensively in CAE, CAD, and CAM. These letters stand for Computer-Aided Engineering, Computer-Aided Design, and, Computer-Aided Manufacturing. They are mostly software rather than hardware techniques. The subject of this book is theory and application of industrial electronics hardware.

1

Electric Circuit Components

IT IS PRESUMED that you have at least a fundamental knowledge of components such as resistors, capacitors, inductors, transistors, and other amplifying devices. These subjects will not be repeated in this book.

Transducers are also called *sensors.* They are necessary in order to exert control—especially feedback control—over an industrial system. Feedback control is also called *servo control,* or *closed-loop control.*

Suppose, for example, a motor that is used in some industrial system must have a controlled speed. By watching the motor and listening to it, you could make manual adjustments to maintain the desired speed. Also, you could observe and listen to the system that is driven by the motor to determine the effect of the motor speed. In this simple closed-loop system, your eyes and ears sense the need for more or less speed. You then make the necessary adjustments.

If you didn't know what the speed was to begin with, it would be very difficult to hold the speed constant using a manual speed control. In other words, you wouldn't know whether to make it go faster or slower if you didn't know its present speed. In fact it may not require any control at all if its state is correct. So, you need a means to sense the speed.

Automatic closed-loop speed control systems use transducers to sense the speed. The electronic circuitry—using the transducer input—then automatically maintains the speed.

As another example, suppose that bottles are being filled on an assembly line. The amount of liquid flowing into the bottle might be set by using only the size of a holding container. Another way is to sense the level of the liquid input so it can be turned off when it has reached the predetermined value. In that case, some kind of sensor is needed to determine the level—that is the job of the transducer. Many sensors used in industrial electronics are electronic components. They will be discussed in this chapter.

In addition to sensors, there are some other important electronic components that should be reviewed, such as the transformer. Transformers will be discussed in a later chapter. Other electric components—such as motors, generators, and relays—are also discussed later in this book.

CHAPTER OBJECTIVES

Following are some of the questions you will find answered in this chapter.

- What is a passive transducer and what are some examples?
- What is an active transducer and what are some examples?
- What basic methods of generating a voltage are used for making transducers?
- What is an LAD used for?
- How is an electric current used to directly cool an area?

TRANSDUCERS

A transducer, or *sensor*, is a component that makes it possible for one type of energy to control another type of energy. The output of a transducer is always a response to the input.

A loudspeaker permits electrical energy to control sound energy; and, a microphone permits sound energy to control electrical energy. So, loudspeakers and microphones are examples of transducers.

You may see transducers defined as being devices that convert energy from one form to another. Although it is convenient to think of then that way, they cannot actually convert energy.

The idea that transducers convert energy is an example of a model. We use it as a convenient way of thinking of a component, circuit, or system. As long as we do not forget that it is a model, and not a precisely correct explanation of action, it is a good way to think of circuit behavior.

Transducers are the sense organs of an industrial control system. They make it possible to measure prevailing conditions so the control system can determine if any adjustment is necessary.

In addition to their use in control systems, transducers are also used for making measurements in many systems. Assume, for example, it is necessary to measure the temperature of an oven. Figure 1-1 shows the control system. A transducer will be used to convert the heat into an electrical signal that is easily measured. The transducer signal is directly related to the amount of oven heat. As shown in the illustration, it can be used for controlling the amount of heat in the oven, or it can be used for measuring that heat.

The control system in Fig. 1-1 employs feedback. If the oven cools, the output of the transducer (and amplifier) is used to increase the input power to raise the heat. On the other hand, if the oven gets too hot, the feedback loop, or circuit, decreases the input power to the oven.

In the system of Fig. 1-1, the transducer is used to measure the furnace heat. Its output signal is directly related to that heat. In transducer circuits the quantity being measured is called the *measurand.* Not all control systems are closed-loop types like the one shown in Fig. 1-1. However, closed-loop circuits are used when very precise control is necessary.

For most measurements and control systems, it is not possible to use the measurand directly. For example, there is no electronic system that can sense heat in a control system. If heat

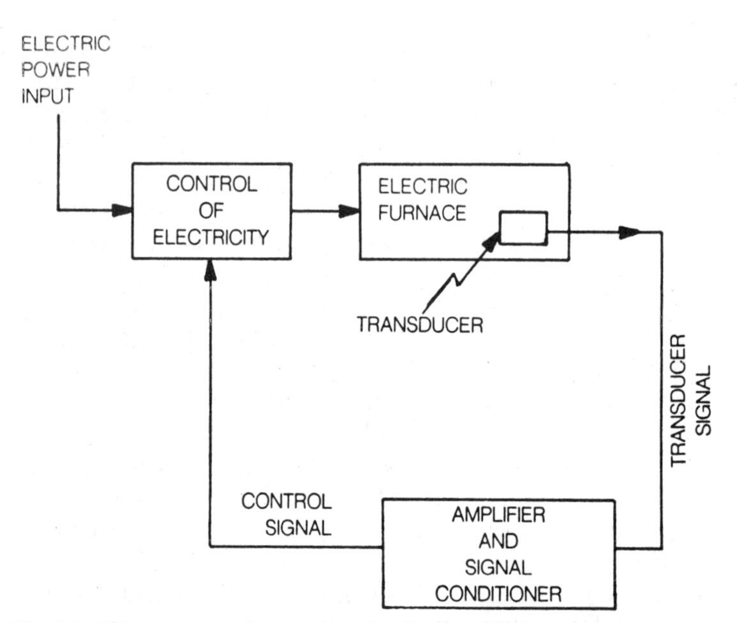

Fig. 1-1. This oven control is an example of a closed-loop control.

is to be regulated, it must be converted to an electrical signal that the electronic regulating system can recognize.

As another example, you might want to continually monitor temperature at a remote mountain top. One way would be to station a person at that position and use a telephone to periodically convey the temperature information. But that would be an unpleasant and unnecessary job because the temperature can be *telemetered* to a more convenient location.

Telemetering means measuring at a distance. In order to accomplish it in this example, it would be necessary to convert the temperature into an electrical signal that could then be transmitted either over the telephone or by radio waves.

All transducers can be classified into two distinct categories. *Active transducers* generate a voltage that is proportional to the measurand. Since there are only a limited number of ways to produce a voltage, it follows that active transducers must use one or more of these basic methods. Reviewing the methods of generating a voltage will be a good introduction to active transducers.

Passive transducers do not generate a voltage. Instead, they produce a change in resistance, capacitance, or inductance that is somehow related to the quantity being sensed. Two types of passive transducers must be considered: *linear* and *nonlinear.* The nonlinear type will be described later in the chapter.

You sometimes see resistors, capacitors, and inductors referred to as *circuit parameters.* A *parameter* can be defined as a value you choose in order to get the answer you want. For example, if you were asked to draw a square that has an area of 4 square inches, the parameter you would choose would be a side. Making the side 2 inches long would give you a 4-inch-square area.

In electricity and electronics, a parameter is something that you choose in order to make the circuit perform the way you want it to. Always remember that parameters can be chosen. For example, if you wanted a certain time constant, the parameter you would choose would be either resistance and capacitance, or resistance and inductance.

If you want a battery to deliver a certain amount of current—that is, the current and the battery voltage are known—the parameter you would choose would be resistance. So, passive transducers perform by changing a circuit parameter.

ACTIVE TRANSDUCERS

The best introduction to active transducers is to review the methods used for generating a voltage. They are listed and discussed in this section.

Chemical Method. One of the oldest methods of generating a voltage is to immerse two dissimilar metals in either an acid or alkali solution. In either solution, a voltage will be produced across the metals.

This method of generating a voltage is often demonstrated in schools by inserting a piece of coat hanger wire (with the paint scrapped off) and a piece of copper wire into a lemon or grapefruit. The acid in the fruit and the dissimilar metals combine to make a very simple battery. The voltage of this battery is easily measured. If you haven't done this experiment, you should try it with a soft drink, as shown in Fig. 1-2, instead of a lemon. You can produce some surprising results.

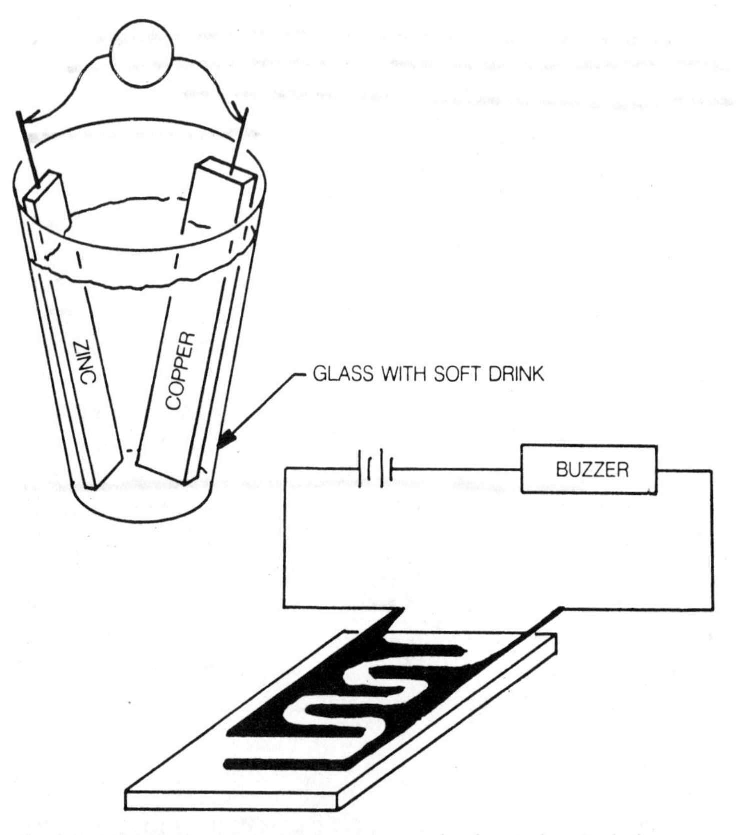

Fig. 1-2. This simple experiment demonstrates the chemical method of generating a voltage by placing two dissimilar metals in a glass of soft drink. Also shown in this illustration is an example of how chemicals can be used to make a transducer, sensing rain. Salt crystals are sprinkled on top of the transducer. When it rains, the salt becomes an electrolyte and completes the circuit between the two conductors, which sets off a buzzer.

The chemical method of producing voltage is not one of the most important for making active transducers, but there have been some examples. You can make a simple device to sense the presence of rain or water, as in a basement by using one of the very simple construction techniques also shown in Fig. 1-2.

If you design a system that has dissimilar metals in contact, and these metals are in an atmosphere or solution that is acidic or alkali, the resulting *galvanic action* can produce corrosion and other undesirable results. This *galavanic action* is due to the current produced by the voltage generated by the chemical method. Although limited as a transducer, the chemical method is important for your understanding of portable systems.

Photoelectric Method. There are certain materials that generate a voltage when exposed to light. An example is germanium. These materials can be used in transducers to sense the presence and amount of light. The voltage produced is proportional to the amount of light present.

Do not confuse the terms *photocell*, which produces a voltage, and *photoresistor*, which produces a change in resistance due to light exposure. They are two different devices. Remember that the photocell is an active device. Both photocells and photoresistors are used as sensors in industrial systems.

An example of photocell use is shown in Fig. 1-3. A light is directed toward a photocell on an assembly line. Every time a box moves in front of the light source, the photocell loses its voltage. That, in turn, represents a pulse input to the counting system. The pulse may have to be shaped so that it is capable of operating the counter system.

You could use a photoresistor in the same application.

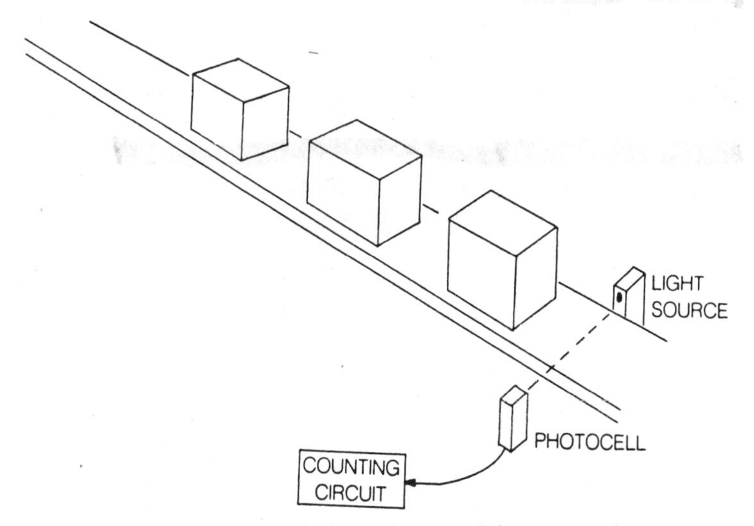

Fig. 1-3. The light source and photocell are used for counting boxes on an assembly line. When a box passes between the two devices, the beam is broken. This produces a pulse, adding one to the counting circuit.

Light falling on the photoresistor changes its resistance value. When a box is present it blocks out the light and the corresponding resistance change is used, after conditioning, to operate the counter. The photocell and the photoresistor can be operated in the system of Fig. 1-3 although one is an active and the other is a passive transducer.

Various devices used in industrial electronics are light activated and are called optoelectronic devices. An example is the Light-Activated Diode (LAD). These devices will be discussed throughout this book.

The photoresistor and photocell discussed in this chapter are not considered in the optoelectronics group. Technically, they are electrical as opposed to electronic components. Photocells are also light-operated devices but they do not fall under any distinct category in the group of transducers being discussed here. Technically, they are photochemical devices that produce an output voltage in proportion to the amount of light that they are exposed to. If you say that all voltages are generated by one of the six methods discussed in this chapter, then the photocell actually cuts across two methods—photoelectric and chemical.

TE Generators. Figure 1-4 shows some important thermoelectric concepts used for making active transducers. For the *Seebeck effect*, two dissimilar metals are heated, which produces a voltage across the ends of the conductors. However, the voltage may not be measureable under normal lab conditions.

Two materials that do produce a measurable voltage are iron and constantan. They are used in a *thermocouple*—a device that generates voltage in proportion to the amount of heat present. Figure 1-5 shows how thermocouples are used for current measurement.

The current being measured flows through a wire and generates heat. A thermocouple, welded to the wire, produces a voltage that is directly related to the heat. Therefore, the voltage is directly related to the current. The voltmeter is calibrated to display the amount of current being measured. This instrument is called a thermocouple ammeter.

The *Peltier effect* was first noticed many years ago but only in this century were materials understood sufficiently to produce significant results. As shown in the illustration, heating one junction causes the other junction to cool. This effect has

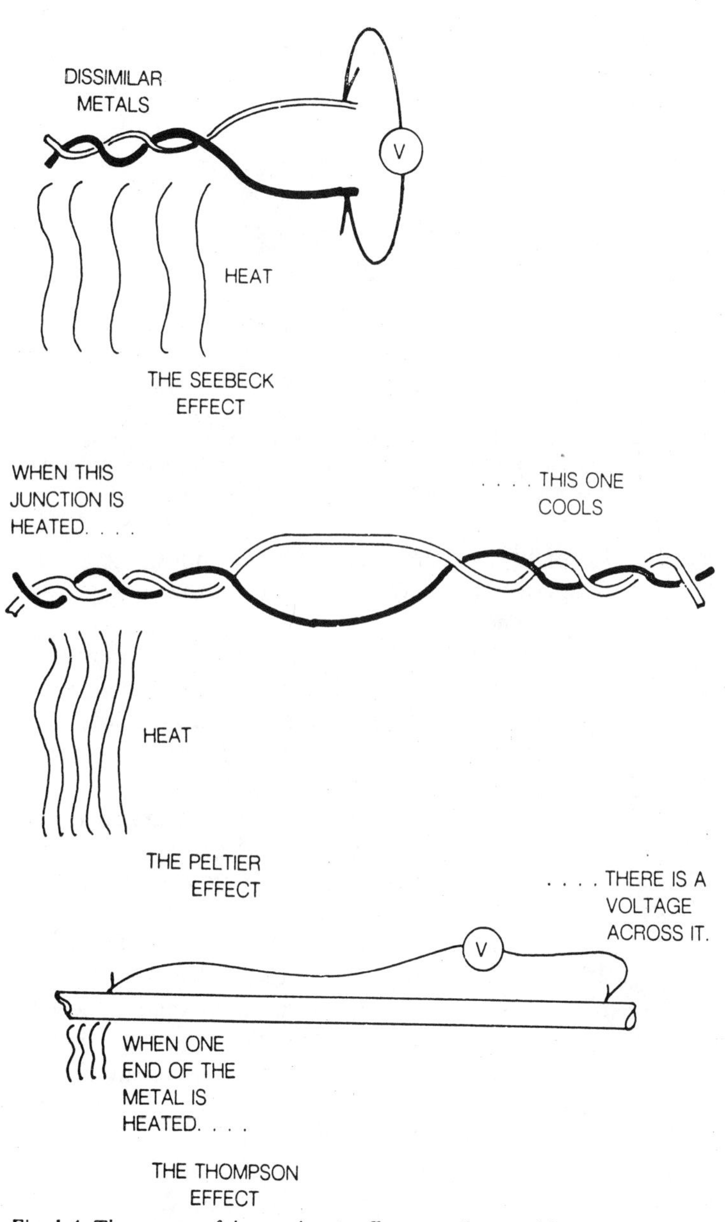

Fig. 1-4. Three types of thermoelectric effects are illustrated here. The Seebeck effect produces a voltage when a junction is heated. The Peltier effect produces a cooling at one junction when the opposite one is heated. The Thompson effect produces a voltage across a rod when one end of the rod is heated.

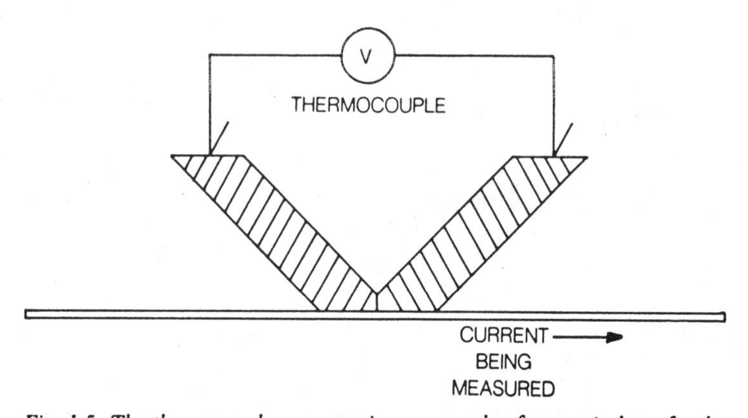

Fig. 1-5. The thermocouple ammeter is an example of a practical use for the Seebeck effect. Here, the thermocouple is against a current carrying wire. The wire is heated by the current to be measured. This produces a thermocouple voltage, calibrated on a scale to display the amount of current.

been used for making small air conditioners and refrigerators. The appliances are not widely distributed but the concept behind them is well known.

The third effect, illustrated in Fig. 1-4, is the Thompson effect. When you heat one end of a bar, as shown in the illustration, a voltage is developed across that bar. Thermocouples are used extensively in devices for measuring high temperatures in furnaces and other very hot environments.

The Piezoelectric Effect. Some materials, such as barium titanate, produce a voltage across their surface when they are under pressure. Two examples of their use are *accelerometers* and *load cells.*

The accelerometer, as the name implies, is used for measuring acceleration. The principle is illustrated in Fig. 1-6. A heavy weight exerts pressure on the piezoelectric material when acceleration takes place. The resulting pressure on the piezoelectric crystal produces a voltage proportional to the amount of acceleration.

This type of measurement is often *telemetered.* In other words the measurement value, in this case a voltage, is converted to a modulated radio wave that is transmitted from the object being accelerated. At the receiver, the radio wave is demodulated and converted into a measurement of acceleration.

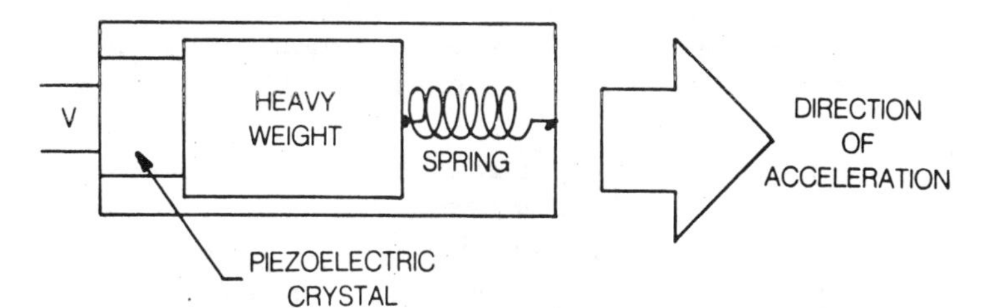

Fig. 1-6. This accelerometer produces a voltage when the heavy weight is thrust against the piezoelectric crystal during acceleration.

A load cell is a method of measuring heavy weights that are exerted by heavy equipment on supports. The load cells are often built into the supports. See Fig. 1-7.

Electromagnetic Effects. Two important concepts are related to electromagnetic behavior: *Faraday's law*, and *Lenz' law*. Faraday's law says that *any time there is relative motion between a conductor and a magnetic field, there will always be a voltage induced in the conductor*. The magnetic field can be cutting across the conductor, or the conductor can be moving through the magnetic field. It is only necessary that there be

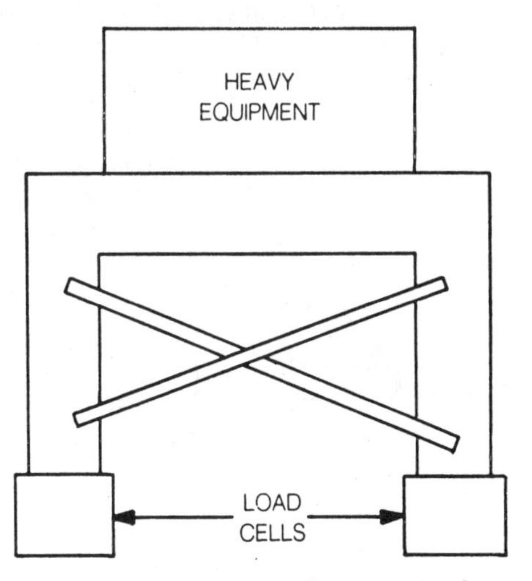

Fig. 1-7. One application of load cells is for measuring heavy weights. The weights can be built into the supports for the heavy equipment. Load cells can also be used to detect vibration in heavy equipment. The same setup is used. Load cells produce a voltage by the piezoelectric effect.

relative motion between them. Relative motion means that if you were standing on one, the other would appear to be moving.

Lenz' law states that *whenever an induced voltage produces a current, the current will have a direction such that its magnetic field will oppose the motion.* You will remember that every time current flows, there are always two effects. A magnetic field is produced around the conductor that carries the current, and there is a voltage drop across the conductor. Faraday's law is used in constructing a mechanical generator or alternator. The generator is made in such a way that there is relative motion between a conductor and magnetic field to produce a voltage.

An alternator works in a similar way. The difference between the alternator and the generator is that the alternator produces an ac voltage. However, that ac voltage can be converted to dc. Alternators are used in cars today, but in the older days dc generators were used.

One practical application of Faraday's law in control systems is the electromechanical tachometer. Basically, a tachometer is a generator that produces a voltage proportional to its speed of rotation. Figure 1-8 shows an example of how it is used. The tachometer shaft is rotated by the wheel. Inside the tachometer is a rotating magnet. The magnet produces a voltage in a coil that is calibrated on the voltmeter as RPM.

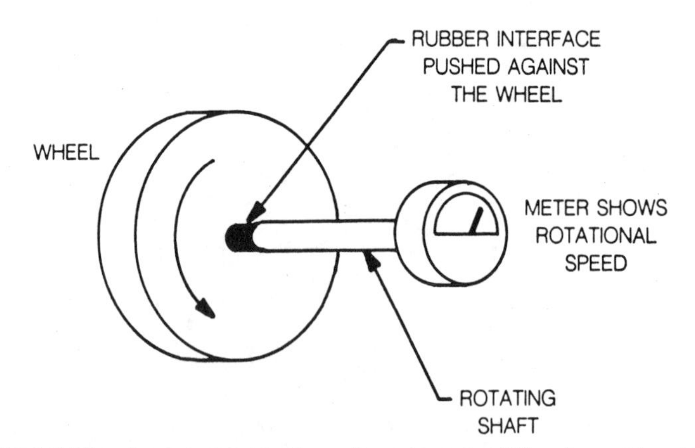

Fig. 1-8. This simple tachometer is a voltage generator. When the rotating shaft turns the generator, a voltage is measured. The voltmeter is calibrated to read RPM.

The speedometer in a car works in a similar way except that the voltage produced is registered as miles-per-hour or kilometers-per-hour. The shaft is connected to the transmission with gears.

Additional Magnetic and Electrostatic Effects. When you magnetize a long bar of iron it becomes shorter after being magnetized than it was before being magnetized. This strange phenomenon is called *magnetostriction.* It is used extensively in very large transducers for ultrasonics. Figure 1-9 shows how it is done.

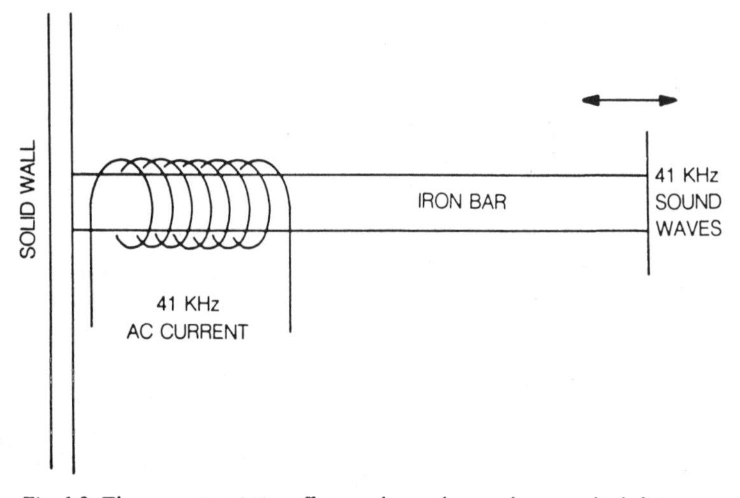

Fig. 1.9. The magnetostrictive effect can be used to produce very high frequency and high amplitude sound waves. The ac current in the coil causes the length of the bar to vary. This, in turn, pushes the flat surface in and out to produce the sound waves. This type of transducer is used to produce high-energy ultrasonic waves in a liquid.

A varying current is passed through a coil which alternately magnetizes and demagnetizes the iron core. That in turn causes its length to vary in accordance with the ac current in the coil. The variation of length is used to force a diaphragm to move back and forth and to produce ultrasonic sound waves. The *villari effect* is closely related to magnetostriction. It states that there is a change in magnetic inductance whenever there is strain parallel to the magnetic field.

Electrostatic Voltage. The last method of generating voltage to be discussed here is called the electrostatic method. It is produced by rubbing two insulating materials together.

When you run a comb made with a nonconducting material through your hair on a dry day, you can produce static electricity. Likewise, scuffing your feet across a carpet, or sliding into a car with plastic seats can produce an electrostatic voltage. However, in practice this method of generating a voltage is not used extensively for making transducers.

Electrostatic generators can produce very high voltages. These voltages are used in x-ray machines to accelerate electrons. The principle of operation for the x-ray machine is shown in Fig. 1-10.

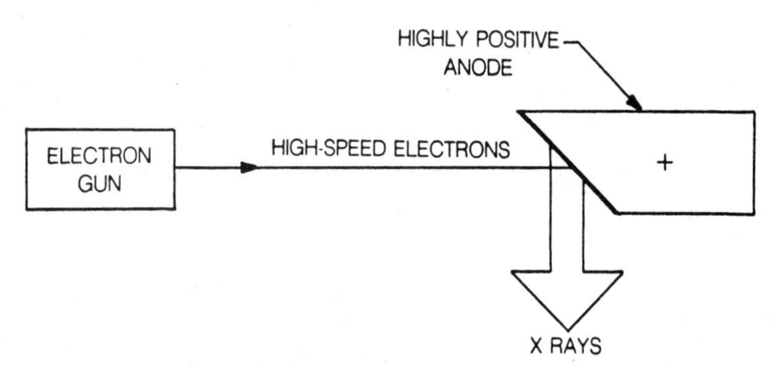

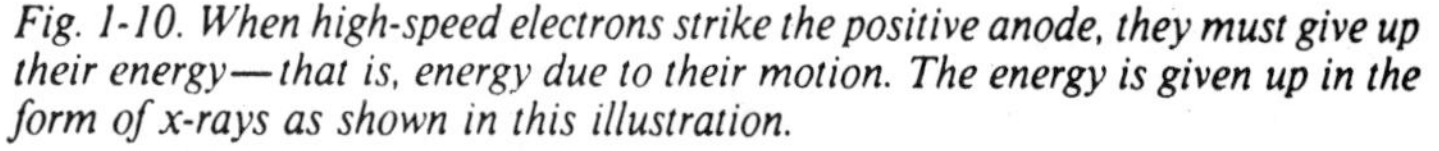
Fig. 1-10. When high-speed electrons strike the positive anode, they must give up their energy—that is, energy due to their motion. The energy is given up in the form of x-rays as shown in this illustration.

An electron gun produces a beam of electrons that is attracted to the highly positive anode. Because of the very positive voltage on the anode, the electrons achieve a very high velocity. That, in turn, means that they have very high energy. When they strike the highly positive anode they come to an abrupt stop. But, they must give up their energy in some form since energy can not be created or destroyed. In this system they give up their energy in the form of x-rays.

These x-rays are used in a number of different industrial operations. They are used extensively in inspection equipment for industrial applications to find internal defects. Electrostatics can also produce some very undesired results. For example, they can destroy some electronic devices.

PASSIVE TRANSDUCERS

Passive transducers usually work by changing resistance, capacitance, or inductance. For that reason you must know how those quantities are affected by changes in dimension or material.

Resistive Transducers. A linear resistor follows Ohm's law. If you double the voltage across it, the current through it will double. However most resistive transducers are nonlinear.

Thermistors. A thermistor is a *therm*ally sensitive res*istor.* It undergoes a large change in resistance value for a relatively small change in temperature.

Thermistors can have a negative or a positive temperature coefficient. With a negative temperature coefficient, there is a decrease in resistance with an increase in temperature. For the positive temperature coefficient, the resistance increases with an increase in temperature. Typical characteristic curves are shown in Fig. 1-11.

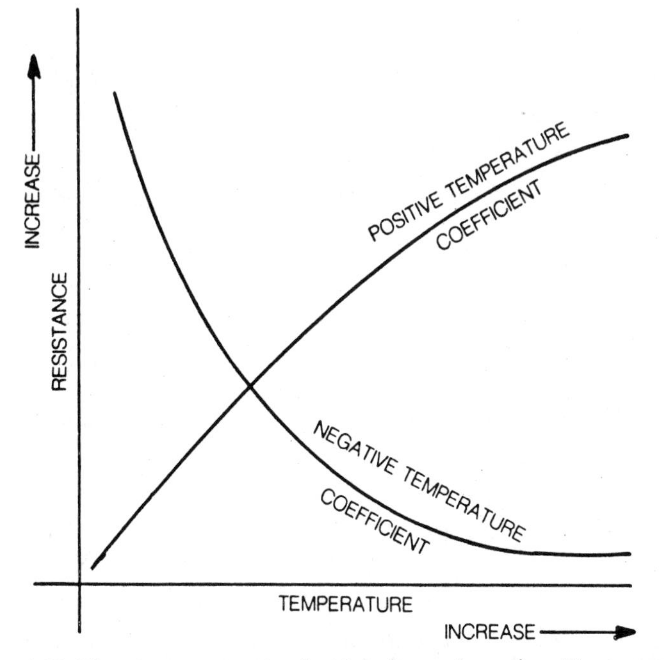

Fig. 1-11. Thermistors are produced with both negative and positive temperature coefficients. Those with negative temperature coefficients are more popular.

Thermistors can be made in very small packages so they produce an almost instantaneous resistance related to a given temperature. Being small, it takes less time to heat the thermistor. This small size is what gives the nearly instantaneous results. For example, a temperature probe is sometimes used by technicians to determine if components are getting excessively hot. The probe has a thermistor in its tip.

However, not all temperature probes are made with thermistors. A thermocouple can also be used for some temperature-measuring applications.

Voltage-Dependent Resistors. The VDR is another kind of resistive transducer. It has a high resistance when the voltage across it is low, and a low resistance when the voltage across it is high.

The most important applications for VDRs are in protective circuits like the one in Figure 1-12. The amplifying device is

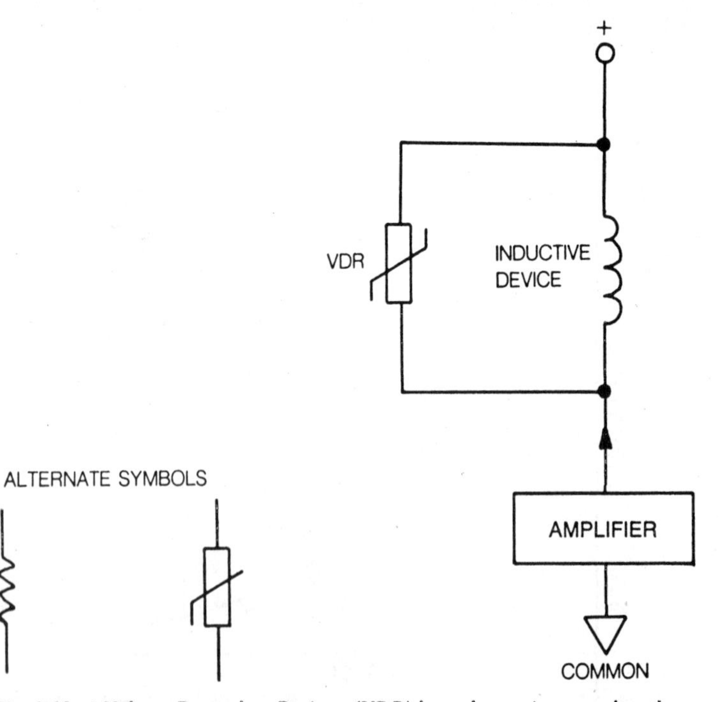

Fig. 1-12. A Voltage Dependent Resistor (VDR) has a low resistance when there is a high voltage across it. In this application, the VDR short circuits the high voltage and prevents it from destroying the amplifier.

used to deliver a pulse to a highly inductive circuit. A very high countervoltage will be generated. In fact, the countervoltage will likely destroy the amplifying device.

The VDR connected across the inductor protects the amplifying device. At the instant the countervoltage is at its maximum value, the VDR resistance is very low. So, it offers a short-circuit path across the coil and kills the countervoltage.

The Strain Gauge. The resistance of a resistor depends upon its length, its cross sectional area and the material from which it is made. The basic equation for resistance of a conductor is:

$$R = \rho \text{ (length}/A)$$

Resistivity (ρ) of the material is a property of the material used for making the conductor. The length of the conductor and its cross sectional area must be given in the same units.

The equation is important because it shows that the resistance can be increased by either increasing the length or decreasing the diameter. It is this property of conductors that is used in making a strain gauge. (See Fig. 1-13.)

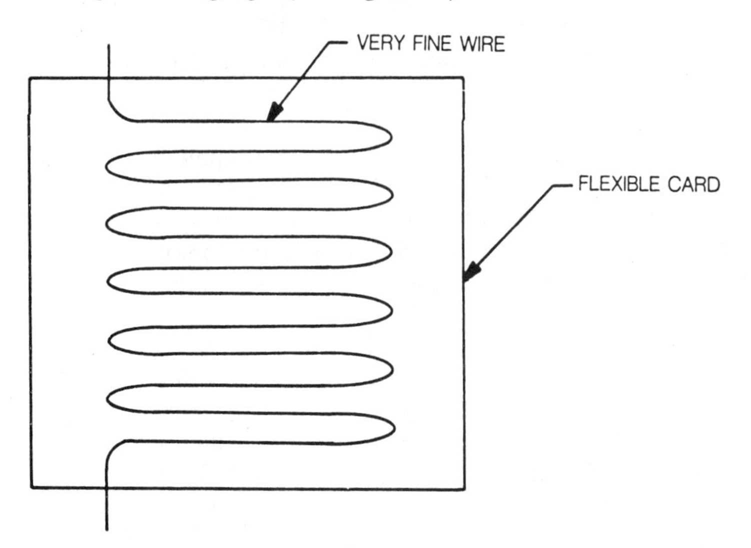

Fig. 1-13. This simple strain gauge is made by bonding a very fine wire to a flexible card. When the card is twisted or bent, the wire is stretched and its cross-section is reduced. The resulting small change in resistance is measured and used as an indication of strain.

Remember that *stress* is a force that produces a deformity, called *strain*. When stress is applied to the strain gauge it causes the wires to stretch, and when they stretch their diameter is decreased. The combined effects of increased length and decreased diameter change the resistance of the strain gauge. The amount of deformity, and hence the amount of resistance change produces a change directly related to the strain.

Capacitive Transducers. A capacitor is a component that stores energy in the form of an electrostatic field between its plates. To understand its operation, it is useful to review the factors, called parameters, that determine capacitance.

Capacitance (C) is a measure of how much energy can be stored in a capacitor. For a simple capacitor, like the one in Fig. 1-14.

$$C = \mathrm{K}\frac{A}{d}$$

Where C = the capacitance of the capacitor.
A = the area of the plates facing each other.
d = the distance between the plates.
K = a constant that depends upon the property of the dielectric.

The value of K in the equation depends upon the *dielectric constant* of the material between the plates. It tells how much more energy can be stored in the capacitor using that material than could be stored if the dielectric was air. Observe, however, that capacitance is not a function of the type of metal used for the plates.

This equation shows that increasing the area of the plates or decreasing the distance between the plates will increase the capacitance. It also shows that capacitance is dependent on the type of insulating material between the plates. Figure 1-14 shows three ways that these parameters can be changed in order to sense a quantity, such as a liquid or a position.

For the *liquid indicator*, the capacitor has an air dielectric when the liquid is low in the tank. However, when the tank is full, the insulating liquid becomes part of the dielectric. So, the

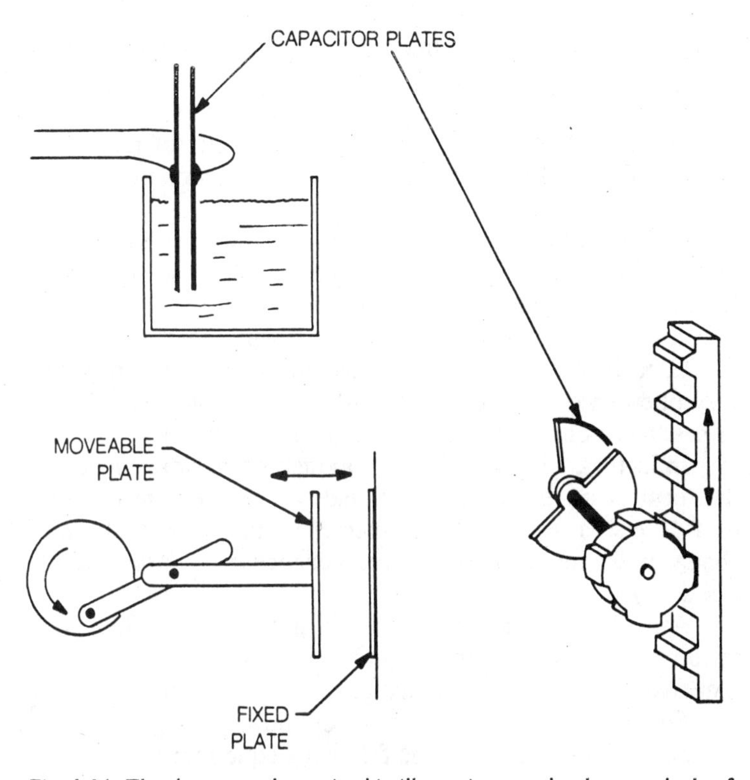

Fig. 1-14. The three transducers in this illustration use the three methods of changing capacitance. In the liquid depth indicator, the type of dielectric is changed. The resulting capacitance change is used to indicate liquid depth. In the second illustration, a variable capacitor is rotated. As the position of the gears changes, change in capacitance is used to indicate position. In the third illustration, a moveable plate causes the capacitance to increase and decrease at a rate fixed by the wheel. That increase and decrease in capacitance is used to cause a variable frequency output.

liquid changes the dielectric constant (K) in the above equation. That, in turn, changes the capacitance.

The effect of change in capacitance affects the electronic circuitry. The capacitive transducer can be used when the liquid level cannot be seen.

In the *motion detector,* the area of the plates facing one another is changed. This is a typical variable capacitor. When the area of the plates is changed, the capacitance is changed.

That, in turn, changes the frequency of a tuned circuit and the frequency of the oscillator.

Two frequencies represent the ends of the travel. They are the highest and lowest frequencies. When either of these frequencies is sensed by the tuned circuit, the direction of travel is reversed.

In the third example of Fig. 1-14, the distance between the plates is changed. That in turn changes the capacitance and frequency of the tuned circuit. Varying capacitance in a tuned circuit produces a frequency modulated signal.

Capacitive transducers are used in ac circuits. They are very often used in tuned circuits where the capacitance determines the resonant frequency of a given LC circuit.

Inductors. Inductors are components that store energy in the form of an electromagnetic field. When they are used as transducers, they are usually connected in such a way that some change in the inductance produces a change in an output signal frequency.

As with the capacitive transducer, inductors are often used in tuned circuits where the frequency is sensed. Any change in that frequency represents a change in the sensed quantity.

For the simple solonoid inductor shown in Fig. 1-15, inductance is determined by the following equation:

$$L = \frac{4\pi AN^2}{l}$$

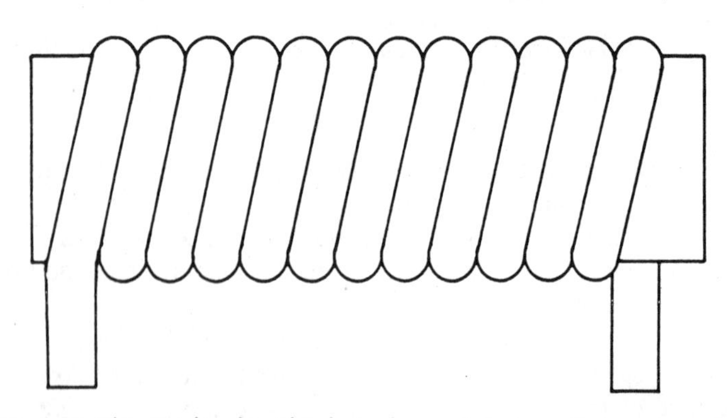

Fig. 1-15. This simple solenoid inductor has an inductance that can be easily computed.

Where L = the inductance of the inductor.
A = the cross-sectional area of the coil.
N^2 = the square of the number of turns of coil wire.
l = the length of the coil.

Figure 1-16 shows variable inductance that can be used for sensing motion. It works by changing the amount of soft iron inserted into the core. The greater the amount of soft iron in the core, the greater the flux leakage, and therefore, the greater the inductance.

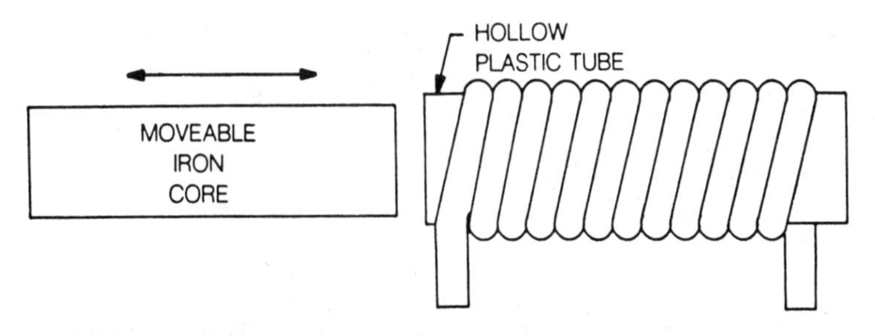

Fig. 1-16. An inductive transducer can be made by moving an iron core into and out of the center of the inductance. The amount of inductance depends upon the amount of iron core in the coil.

Another method of varying inductance is to short-circuit some of the turns. It is not a common practice to change the diameter of the coil in transducers.

The Bridge Connection. Passive transducers are usually connected in a bridge circuit like the one shown in Fig. 1-17. This bridge can be supplied with either an ac or dc voltage.

One reason for using the bridge is to make the measurement independent of the power supply voltage. This is especially important when the power supply is a battery. When a battery is delivering current, its voltage decreases over a period of time.

Note that two transducers are used in the bridge. One is called the *control* transducer and the other is called the *measuring* transducer. These transducers must be very carefully matched so that when they are exposed to room temperature, or any other temperature, their resistance values will be exactly the

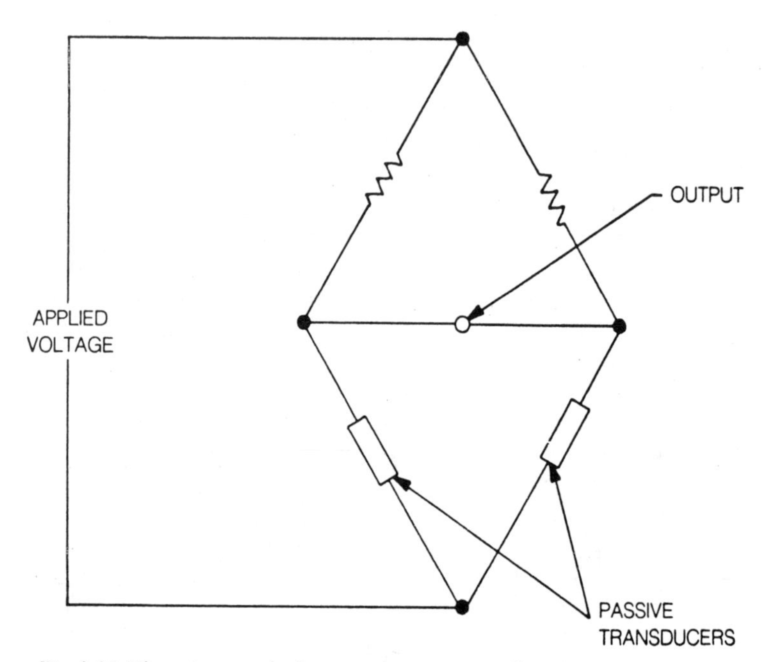

Fig. 1-17. Thermistors and other transducers are usually connected into a bridge circuit like the one shown here.

same. That will prevent the ambient (surrounding) temperature from affecting the bridge output.

The measuring transducer is placed where it will sense some quantity. This causes an unbalance in the bridge, and current to flow through the center leg. A current-measuring instrument in that center leg can be calibrated to display a measurand. For example, if this is a bridge used for measuring temperature, the meter in the center leg would be calibrated into degrees Celsius or degrees Fahrenheit.

With identical transducers in the legs, the same amount of power supply current flows through each. Therefore, the heating effect of the supply current will be the same for each transducer. Each leg will be affected in the same way; so, their effects are balanced. Saying it another way, the heating effect of the power supply current will not affect the bridge output.

To summarize, there are three advantages for using a bridge circuit for transducers:

- The condition of the power supply, and its terminal voltage will not affect the use of the transducer.
- Changes in ambient temperature do not affect the balance of the bridge. Only the difference in the temperatures of the transducers will affect the current through the center leg.
- Power supply current flowing through the transducers affects both in the same way, so, it is not a factor in determining the center-leg current.

SUMMARY

One of the most important electrical components used in industrial electronics is the sensor, or transducer—the terms are used interchangeably.

Transducers do not convert energy from one form to another. Instead, they permit the energy of one system to control the energy of another. A microphone, for example, permits sound energy to control electrical energy. In that way, it is a transducer. Loud speakers, on the other hand, permit electrical energy to control sound energy; therefore, they are also transducers.

Transducers perform the sensing necessary for controlling industrial processes and machines. They are the eyes and ears of the electronic industry. In addition to seeing and hearing, they also perform such jobs as touching or even smelling, as in the case of poisonous gases. So, transducers are not used exclusively in industrial electronics; but, they are one of the most important components in control systems.

Transducers are also necessary for making measurements. The measured quantity is called the *measurand.* When measurements are made at a distance the system is *telemetering.*

Transducers can be divided into two main classes: active and passive. Active transducers generate a voltage that is in some way related to the measurand. Passive transducers change the resistance, capacitance, or inductance of a circuit in accordance with the measurand.

Active transducers use one of six methods of generating a voltage. As an example, photocells produce a voltage that is dependent upon the amount of light they receive. They use the photoelectric method of generating a voltage.

Passive transducers operate by changing one of the determining factors of resistance, capacitance, or inductance. For example, the capacitance of a capacitor depends upon the distance between its plates. That distance can be varied in accordance with some measurand.

It is common practice to connect passive transducers into bridge circuits. That way they are not dependent upon the applied voltage, ambient temperature conditions, or the amount of power supply current flowing through them. The bridge circuit requires two transducers—one being a perfect match of the other. Any changes in current, voltage, or ambient conditions will change both in exactly the same amount. That way, those changes do not affect the reading or the measurement.

SELF TEST

1. When you magnetize an iron bar, it gets:
   (A) shorter.
   (B) longer.

2. In the setup of Fig. 1-18, the light strikes the transducer each time a slot is present. In this control system, the

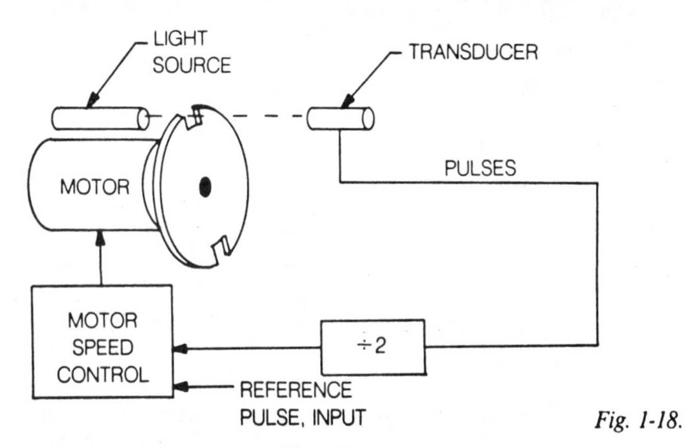

Fig. 1-18.

   output of the transducer is used to control the speed of the motor. Which of the following statements is correct?
   (A) The transducer must be active.
   (B) The transducer must be passive.
   (C) Either statement is correct.

3. In a certain circuit a variable resistor is used to adjust the voltage. Is the variable resistor an active transducer?
   (A) Yes.
   (B) No.

4. A resistor that undergoes a wide change of resistance values for a relatively small change in temperature is called:
   (A) a TE generator.
   (B) a thermistor.

5. A certain hand-cranked generator turns easily when there is no load. Shorting the output terminals creates a heavy

load and the generator is hard to turn. This is a demonstration of:

(A) Lenz' law.
(B) the Peltier effect.

6. Galvanic action is produced by the:
   (A) chemical method of generating a voltage.
   (B) photoelectric method of generating a voltage.

7. The metals in contact in Fig. 1-19 are called *faying surfaces*. They:

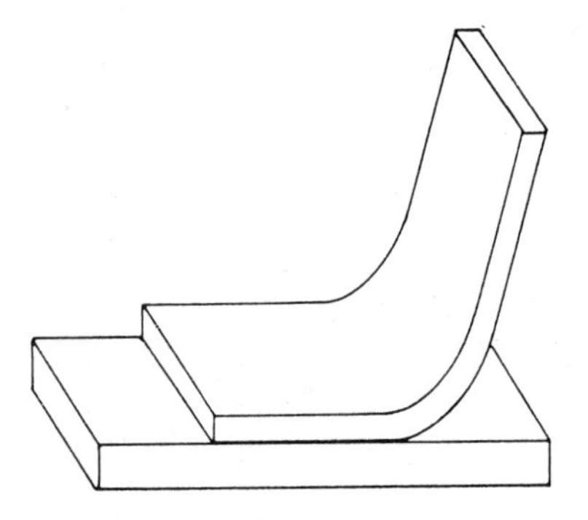

Fig. 1-19.

   (A) should only be used when the surrounding air is acid or alkali.
   (B) should never be used in an acid or alkali atmosphere.

8. Barium titanate is used for making:
   (A) light-activated devices.
   (B) chemical transducers.
   (C) Neither choice is correct.
   (D) Both choices are correct.

9. Is the following statement correct? Solar cells are photoelectric components.
   (A) Not correct.
   (B) Correct.

10. A load cell is:
   (A) a type of battery.
   (B) a type of transducer.

ANSWERS TO SELF TEST

1. (A) This is known as *magnetostriction*. This characteristic is utilized in some types of electricity to sound transducers, especially those used in liquids.

2. (C) Either an active transducer or passive transducer can be used in this application. The active light transducer is called a photocell, whereas the passive light transducer is called a photoresistor. The pulses from the sensor are used to sense the speed of the motor. That speed, in the form of pulses, is compared with a reference pulse frequency. If there is any difference, a correction is made in the motor speed. This system is called the phase-locked loop and it is discussed in another chapter.

3. (B) No, the variable resistor is not a transducer. It does not permit one type of energy to control another type. It simply controls electrical energy.

4. (B) Thermistors are made with positive and negative temperature coefficients. The negative temperature coefficient type is more popular.

5. (A) Current set up by the generated voltage produces a magnetic field that opposes the motion of the crank. The term *load* in electrical and electronic circuits should not be confused with the term *load resistance*. Technically, load is a reference to how much current a power supply must deliver. When you say a power supply is under heavy load you mean it is delivering a high current. Load resistance is the opposition to current flow from a generator. A heavy load is created when the load resistance is low.

6. (A) Galvanic action can produce destructive effects in faying surfaces. It is an important consideration in the choice of materials used in fabricated equipment.

7. (B) For the same reason discussed in the answer for question #6, materials used for faying surfaces must be carefully chosen and the atmosphere that they work in must also be taken into consideration. Keep in mind that any time two dissimilar metals are placed in an alkali or acid environment a

voltage will be produced. That voltage can cause destructive corrosion in the faying surfaces.

8. (C) Barium titanate is used for making piezoelectric devices. It is a crystalline material that can produce a very large voltage for a relatively small pressure.

9. (A) Solar cells use an electrochemical method of generating a voltage.

10. (B) Load cells are used for sensing pressure that is produced by weight.

2

Electronic Circuit Components

IN THIS CHAPTER you will review two-terminal and three-terminal linear devices. The first two-terminal devices of interest are vacuum diodes. At first you may feel that this is obsolete material; however, magnetrons are examples of vacuum diodes. Also, since there are many vacuum tube diodes still in existence, it would be good to briefly review their operation. Gaseous diodes are still being used in very high-current rectifier systems. Their principle of operation is also reviewed here. Of course, semiconductor diodes are also reviewed in this chapter.

Three-terminal linear devices are used for amplification. Vacuum linear devices are briefly reviewed, because, like the vacuum tube diodes, they are used extensively in electronic equipment even today. The cathode-ray tube is an example of a three-terminal device. Remember that the three terminals do not include the dc operating electrodes necessary for device operation.

Bipolar transistors have, for the most, replaced vacuum tubes in the newer designs. Field effect transistors are becoming increasingly popular because of their low standby power requirements.

Certainly a book on industrial electronics should reflect the newest devices. However, industrial electronic equipment can be very expensive; it is not uncommon for a smaller piece of equipment to cost over $20,000. Manufacturers are not prone to throw that equipment away simply because there are newer electronic devices on the market. As a technician you must be able to troubleshoot all types of electronic equipment using older and newer technology.

CHAPTER OBJECTIVES

Here are some of the questions you will find answered in this chapter:

- What is avalanching? (Avalanching is a characteristic of both tube and semiconductor devices.)
- How does a magnetron work?
- What types of bipolar transistors are used in linear systems?
- What is the difference between an enhancement and a depletion MOSFET?
- How do JFETs work?

In the traditional manner of studying any electronic subject, components are discussed first. Then, circuits which are combinations of components follow. Finally, complete systems round out the study.

To be specialized in industrial electronics, you have, no doubt, studied the components and basic circuits. So, the material in this chapter will be a review. However, there are some specialized components that you may not have had in your studies. Also, some applications are included. Some subjects, now considered to be outdated, are not usually included in most general electronics courses. As an example, you may not have studied vacuum tubes at all, or if you did, it was an adjunct to the study of field-effect transistors.

There are still many vacuum tube systems in operation. They include expensive test equipment, field strength meters, and some control systems. As a technician you might be expected to repair that equipment. Therefore, you should have a basic understanding of how a tube works. You will note that the dc operation of a vacuum tube is very similar to the operation of FETs. A brief review of vacuum tube principles is included in this chapter.

You will review the bipolar and field-effect transistors, with a few special references to applications. An example is the configuration for three terminal devices.

DIODES

At one time you could study diodes in a few pages of a book, and those few pages would cover every aspect of available diodes. Today, some types of diodes have become very complicated devices. They are not only used for *rectifying* and *detecting*—that is, allowing current to flow in one direction and not in the opposite direction—but they are also used for emitting light, producing voltage regulation, capacitance, switching, and many other applications. In some cases, specialized construction of the diode is necessary to achieve those results.

Vacuum Tube Diodes. The earliest diode of any significance was the vacuum tube diode. In England it is called the *Fleming valve*, or just a *valve*. Its operation is very simple. This discussion will explain how electrons are emitted from a surface in various operating devices.

Maybe you think of the vacuum tube diode as being totally obsolete. If so, you may be surprised to know that *magnetrons* and *x-ray tubes* are both examples of vacuum tube diodes that are still being used extensively.

Other Types of Diodes. In the early days of electronics most electronic devices operated by controlling the electron flow in an enclosed region. The diode tube is an example. It is illustrated in Fig. 2-1.

Actually, diode tubes are not constructed as shown by the symbol. Instead, the cathode is at the center of a concentric plate. The inset of the illustration shows that construction, but the symbol works best for discussing how the device works.

The first step is to put some electrons in motion. Then, the motion of the electrons is controlled in some way. That is the basis of tube operation. But then, that is also the basis for the operation of semiconductor devices.

For the vacuum tube diode of Fig. 2-1, electrons are emit-

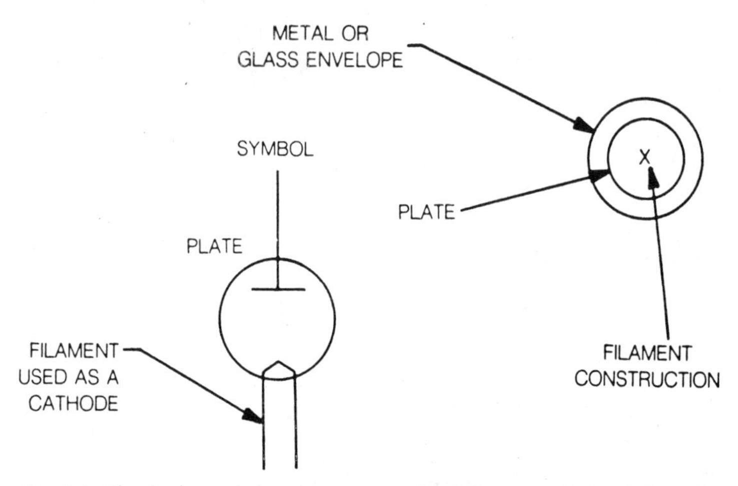

Fig. 2-1. The diode symbol and its construction. Electron emission is from the filament, so this is called a directly-heated cathode.

ted by the cathode. They are attained by the positive voltage on the plate.

The mechanism of electrons flowing from cathode to plate is simple to understand, but it is slightly complicated by the fact that in the real device a cloud of electrons surrounds the cathode. This cloud of electrons is called the *space charge.* Actually then, the current that reaches the plate is due to electrons that are attracted from the space charge.

Getting Electrons. You can consider the operation of a vacuum tube device to be accomplished in these steps:

First, get some electrons.
Second, control their motion.
Third, use circuitry to make use of that control.

There are four ways to get electron emission. You should understand how this is accomplished. The four methods are illustrated in Fig. 2-2.

One method of making a surface emit electrons is to produce a high temperature at that surface. The electrons in the material then achieve a sufficient amount of energy to escape from the surface. This method is called *thermionic emission.* It is used to get the electron emission for the diode of Fig. 2-1.

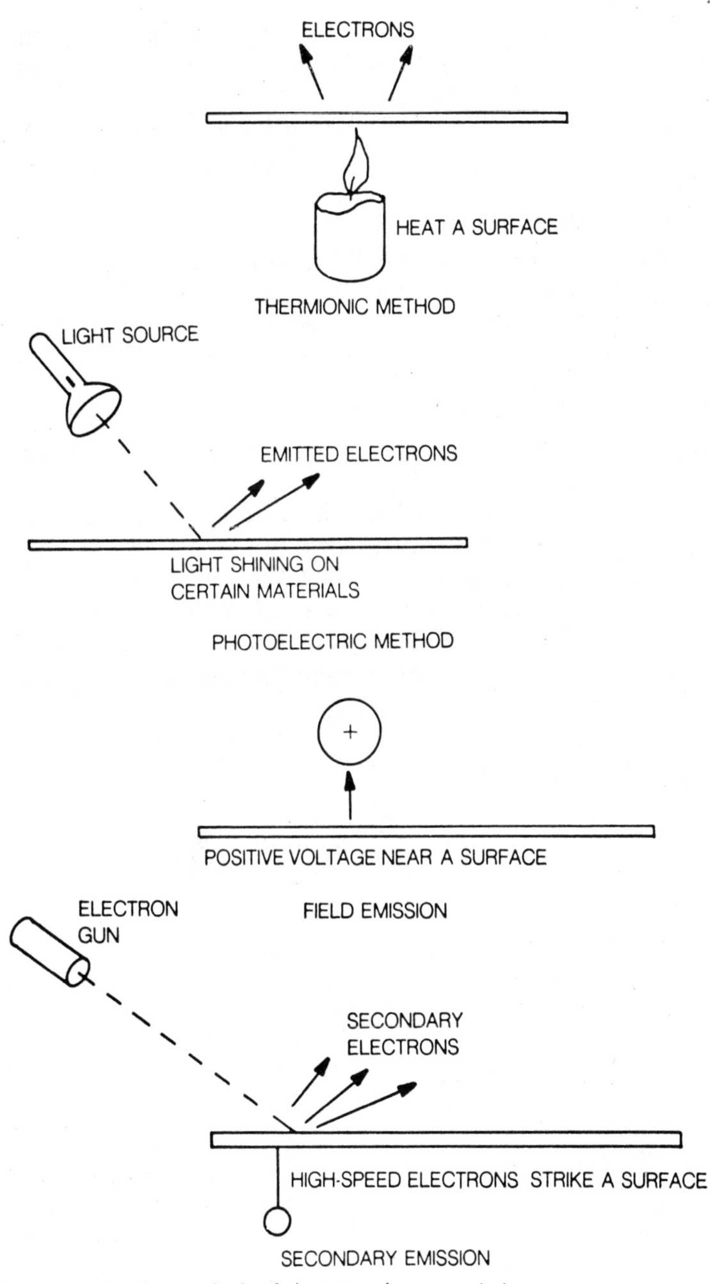

Fig. 2-2. The four methods of obtaining electron emission.

Another way to produce emission is to shine a light on certain materials, such as the materials used in a photocell. This method is called *photoemission.*

Another method of achieving emission is to produce a high positive voltage near the surface of the material, and "drag" the electrons off the surface. This is the method used in the neon tube. It is called *field emission.*

Another emission method is to direct a high-energy beam of electrons onto the surface of the material. The high-energy electrons knock other electrons loose from the surface. This type of emission, called *secondary emission,* is used in photomultipliers.

By far the most popular method of emitting electrons is thermionic, but there are still devices in operation using other types.

The Indirectly Heated Cathode. The diode in Fig. 2-3 is slightly different from the one shown in Fig. 2-1. In this case, the electrons are emitted from a metal sleeve surrounding the filament. The metal sleeve is called a *cathode.* It is coated with materials that readily emit electrons when heated. However, these materials cannot be used directly for making filaments for the direct-heated diode of Fig. 2-1.

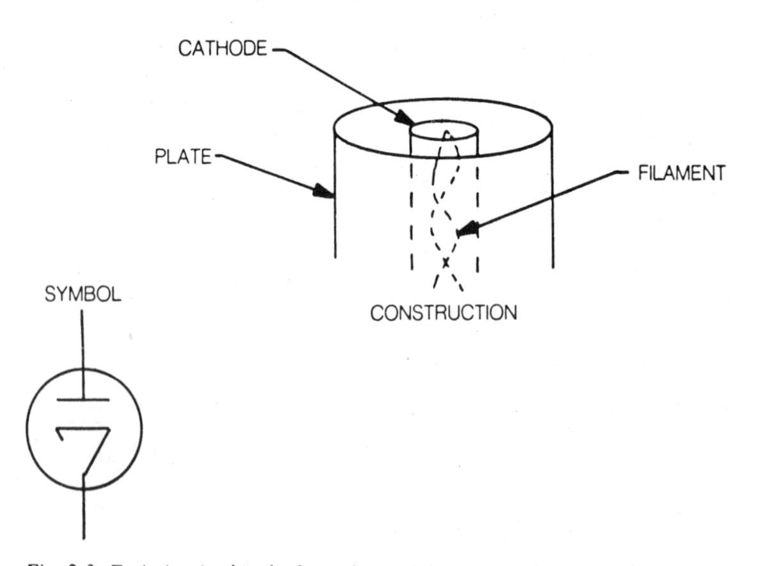

Fig. 2-3. Emission in this diode is obtained from a filament-heated cathode.

The mechanism of current flow in the two types of devices is the same. When the plate is made positive, electrons from the space charge collect around the cathode. They are attracted to the positive plate, and electron current flows through the diode.

When the plate is negative, no current flows through the device. As this characteristic of two electrode tubes was discovered many years ago by Thomas Edison, it is called the *Edison effect*.

Sir Ambrose Fleming of England was first to realize that this characteristic could be used in making the tube a rectifier, or "valve", to convert alternating current to a pulsating direct current. In his application, the alternating current was a signal used in radio communications. The simplified illustration of Fig. 2-4 shows how the diode converts the ac signal input to a pulsating dc output. In this circuit, a current can flow through the diode only on a half cycle when the plate is positive with respect to the cathode. This same unilateral flow takes place in semiconductor diodes, except that it is accomplished by a different method.

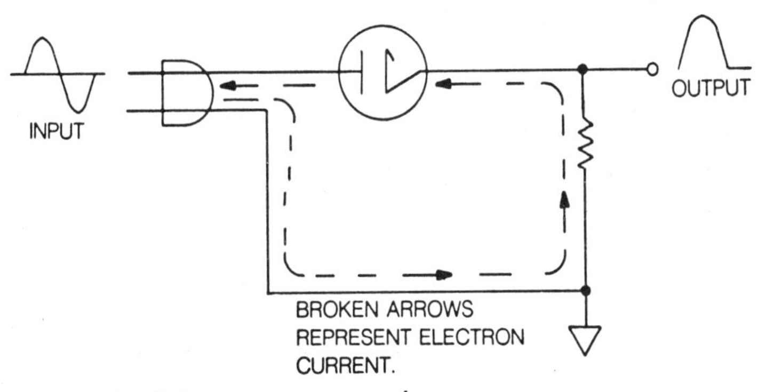

Fig. 2-4. The diode acts as a one-way valve.

A special type of diode called the *phanatron* is illustrated in Fig. 2-5. The dot inside the envelope of a tube always means that an inert gas has been inserted into the envelope of the tube. The presence of gas changes the current flow through the gas tube by a mechanism called *avalanching*. This same mechanism occurs in semiconductor devices as illustrated in Fig. 2-6.

Here, an electron strikes an atom of gas and knocks two

Fig. 2-5. Gas diodes are represented by this symbol. They are still being used for very high-current applications.

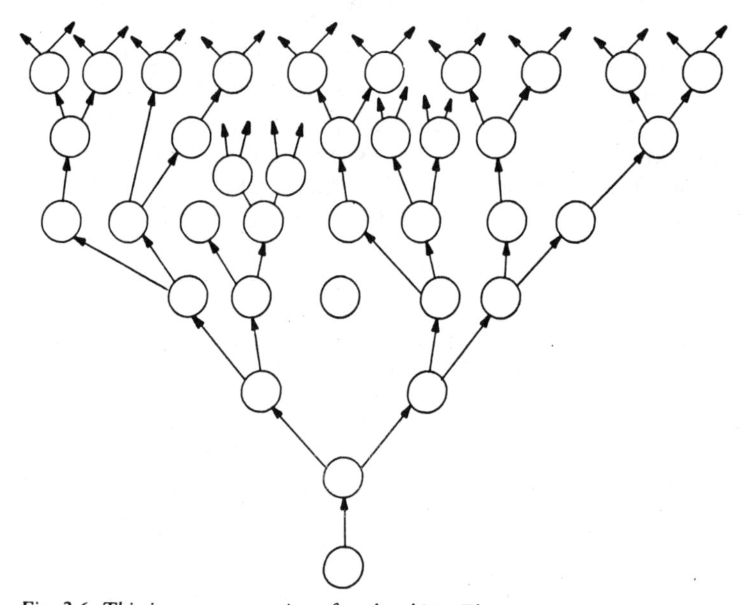

Fig. 2-6. This is a representation of avalanching. The arrows represent electron motion, and the circles represent atoms or molecules.

electrons loose. Those electrons, in turn, strike atoms and knock additional electrons loose. Throughout the device this collision of electrons with atoms increases the flow of electrons in the tube. For each electron, there is a positively charged atom called a *pion* that moves toward the negative cathode where it becomes neutralized in the charge space.

The X-Ray Tube. X-rays are used extensively in industrial electronics for testing materials and devices. Because of their high penetrating ability, x-rays permit us to see inside the device.

The theory of x-ray tube operation was discussed in chapter 1. See Fig. 1-10. The electron gun in the illustration is the cathode. In practice, there are baffles with holes to direct the electrons to the *anode*, or *plate*. Be sure you understand the theory of operation for the x-ray tube. If necessary, read the discussion of Fig. 1-10 again.

If you are using x-ray equipment in your work, be sure to observe the safety rules! You will be entirely safe and protected if you follow the rules.

The Magnetron. Magnetrons are very specialized diodes. They are used as oscillators in ultrahigh-frequency operation. Figure 2-7 shows their construction.

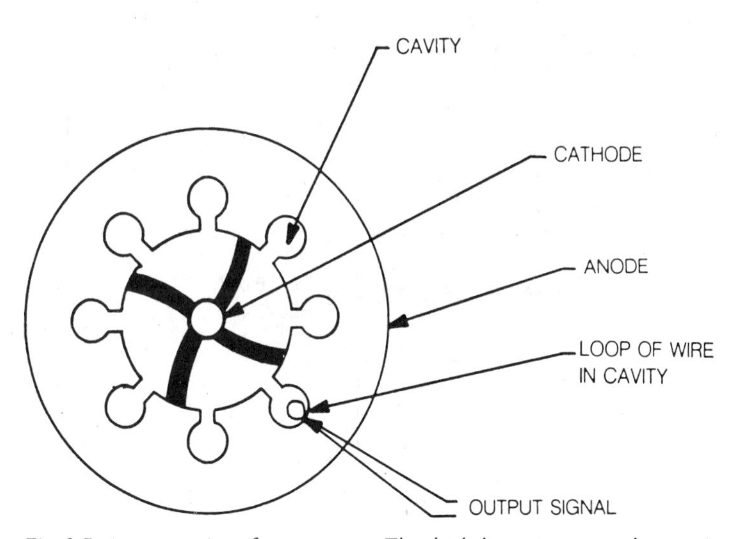

Fig. 2-7. A cross section of a magnetron. The shaded area represent electrons in motion around the cathode.

Magnetrons have a cathode at the center and *resonant cavities* in the anode. A resonant cavity is simply a chamber that can sustain oscillation at a very high frequency. It is roughly equivalent to an LC tank circuit. That is the best way to think of the magnetron.

Electrons are emitted from the cathode, but they do not go directly to the positive plate. Instead, magnetic fields steer them so that they move in a circular fashion around the space between the cathode and the anode.

The electrons tend to bunch together, and their magnetic fields combine. The fields inject energy into the cavities as they move past the cavities.

Think of the cavities as being LC circuits and the moving magnetic fields inducing the voltage into the LC coil. That starts oscillation. A small loop of wire is then used to take the microwave energy away from one of the cavities.

Silicon Diodes. Like the vacuum tube diode, a silicon diode is a unilateral device. In other words, it conducts from cathode to anode, but not in the reverse direction.

The model of a silicon diode is shown to Fig. 2-8. This model is used to explain how current flows through the device. Electron current is assumed in this discussion.

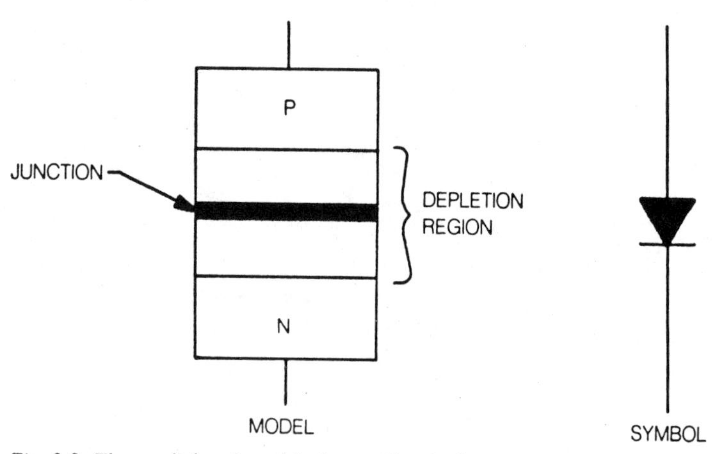

Fig. 2-8. The symbol and model of a rectifier diode.

Electrons are readily available as charge carriers in the N-type material. You should not think of this as being negatively-charged material. It is simply material that can easily conduct electrons. In the P-type material, the charge carriers are not electrons, but holes. Technically, a hole is a place where an electron can go. Electrons flow through the P-type material from hole to hole, but the hole is considered to be the current carrier.

The concept of holes is easily understood from the model in Fig. 2-9. Here balls are moved from left to right, one ball at a time. Note that as the ball moves from left to right, the hole moves from right to left. In a similar way, conduction in P-type material is considered to be by holes that move in the opposite direction to the electrons.

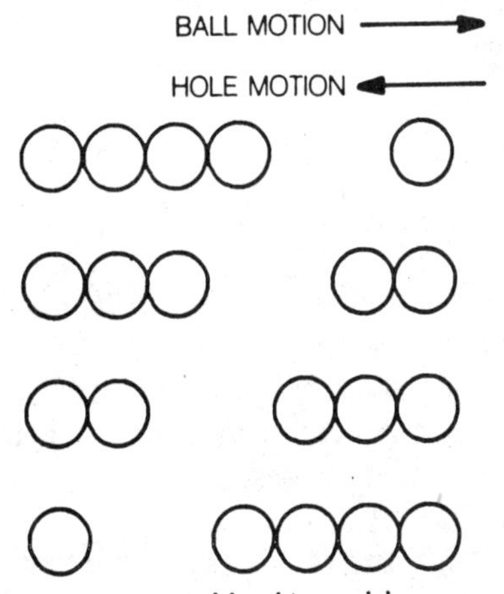

Fig. 2-9. Hole flow is represented by this model.

The overall result is that electrons move through the N-type material and through a "no-man's land" called the *depletion region* into the P material. From that point they move on to the anode from hole to hole.

In the depletion region, there are no surplus charge carriers or, to be exact, there is a very limited number of charge carriers available for current in this region.

The depletion region changes in size according to the voltage across the diode. Three cases are shown in Fig. 2-10. When there is no voltage across the diode, the depletion region is fixed at a size determined during the manufacture of the device. When the diode is forward biased, the depletion region decreases, making it easier for electrons to pass from the N-type to the P-type material. So, current flows through the diode readily

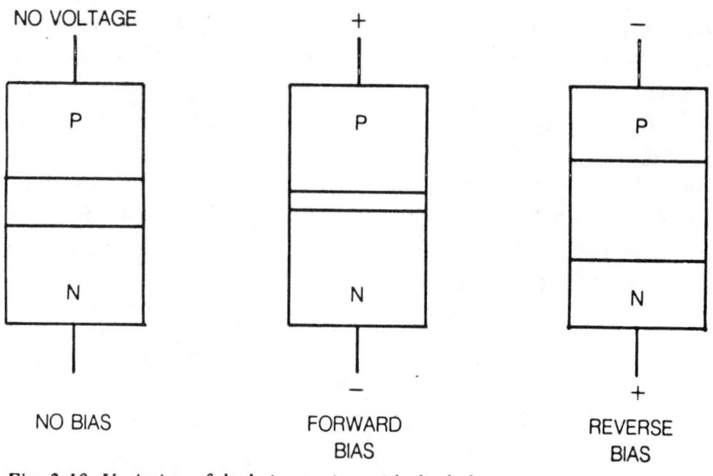

Fig. 2-10. Variation of depletion region with diode bias.

when it is *forward biased*—that is, when its anode is positive with respect to its cathode.

When a diode is *reverse biased*, the depletion region is increased. That makes it nearly impossible for an electron to move through the depletion region from the N-type to the P-type material.

From what has just been said you can see that electrons will flow readily through the diode when the anode is positive with respect to the cathode; but, the diode will not conduct if the anode is negative with respect to the cathode. In this sense its operation is similar to the vacuum tube.

The symbol for the semiconductor diode shown in Fig. 2-8 is important. It consists of an arrow representing the anode and a straight line representing the cathode. Traditionally on all semiconductor devices, the current is assumed to be conventional current flow. The arrows in the semiconductor devices always point in the direction of conventional current (opposite of electron current). Also, the arrows always point toward an N-type material. So, in the semiconductor symbol, the arrow is pointing to the N-type material in the cathode, and it points in the direction of conventional current flow.

You can substitute a semiconductor diode directly for a vacuum tube diode rectifier. This assumes that the specifications of the diode make it suitable for use in that application.

Optoelectronic Diodes. There are two additional important diodes that should be mentioned. They are covered in chapter 3. One is the *light-emitting diode, (LED)*. The LED emits a characteristic red, green, or yellow light when current flows through it in the forward direction.

The second type is the *LAD (Light Activated Diode)*. In the absence of light this diode has a high resistance in the forward direction. However, when it is exposed to light, the forward resistance drops to a low value and the diode conducts with very little resistance.

THREE-TERMINAL ELECTRONIC DEVICES

In 1908, when Lee DeForest connected a grid between the cathode and plate of a diode he opened the electronic era. Since then electronics has become a continually expanding industry. We will take a quick look at the "audion," as DeForest called the triode. Its characteristics are very similar to that of a JFET which we will also discuss in this section.

The devices discussed in this section are sometimes referred to as *analog* or *linear*. The term linear needs to be explained. It has different uses in different applications of industry.

If you are talking about a linear resistor it means that the resistor follows Ohm's law. If you double the voltage across the resistor, the current through it will double, provided it is operated within the limitations set by the manufacturer.

When you are talking about a linear amplifier on a linear amplifying device, it has nothing to do with Ohm's law. Specifically, a linear amplifier is one that has a continuous change in its output for a continuous change in the input. Stated another way, if you apply a sine wave to the input, you get an approximate sine wave in the output.

The linear devices discussed in this section can be operated in the other way, that is, in the nonlinear way. But, for all practical purposes they are classified in the group called linear or analog devices.

All linear amplifying devices are similar in that they are three-terminal devices. That means that one terminal must be used for the input signal and one for the output signal, and one must be common to the input and output.

You can connect these devices in any of three ways. The difference between the devices is the mechanism by which control is exerted over the flow of electric current through them. Usually, we will discuss them in terms of flow of electrons or holes.

Remember that all of these devices operate from a dc power supply. They cannot supply their own power. If they are used as power amplifiers, it means that they are able to control the power supply output.

Likewise, they cannot supply voltage. If they are used as voltage amplifiers, they are simply controlling the power supply voltage that is delivered to them.

Summarizing, they are neither power or voltage amplifiers without a power supply. In their operation they always control the power supply output.

The Vacuum Tube Triode. Figure 2-11 shows the symbol and the construction of a vacuum tube triode. The symbol will be used in this discussion.

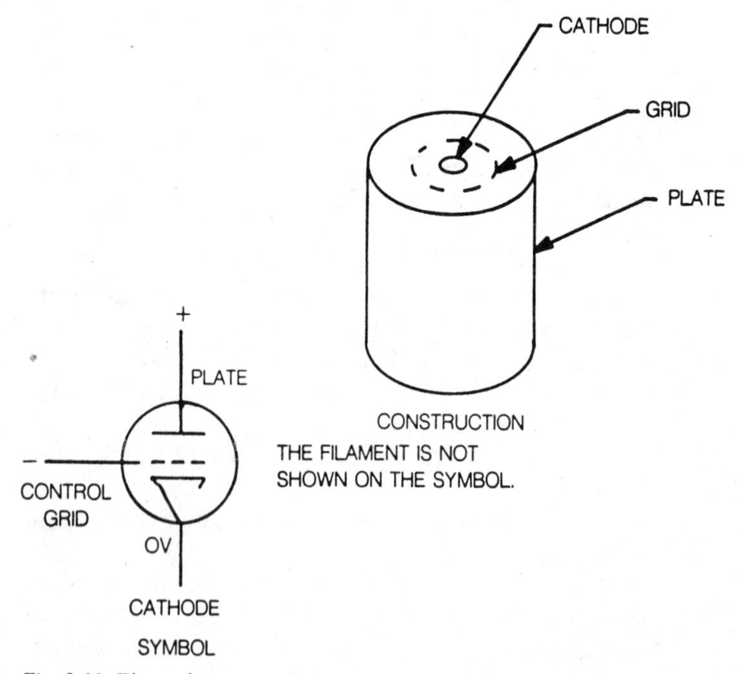

Fig. 2-11. The audion or triode has this symbol and construction.

Electrons are emitted by the cathode and move toward the positive plate. In doing so, they move between the holes in the negative grid. The amount of negative grid voltage determines how many electrons actually reach the plate. The grid can be made so negative that no electrons reach the plate. That condition is called *cutoff.*

The key to understand the operation of this device is knowing that a *relatively small change in control grid voltage can produce a relatively large change in the amount of plate current.* The *plate current* is simply the amount of current that flows through the tube.

Triodes have an important disadvantage. The control grid is made of metal and the plate is made of metal. They are separated by an insulating material (vacuum), so they form a capacitor. In very high frequency operations, the capacitance between the grid and plate allows the signal to "sneak" through the tube without being amplified. The triode is modified for high-frequency operation as shown in Fig. 2-12.

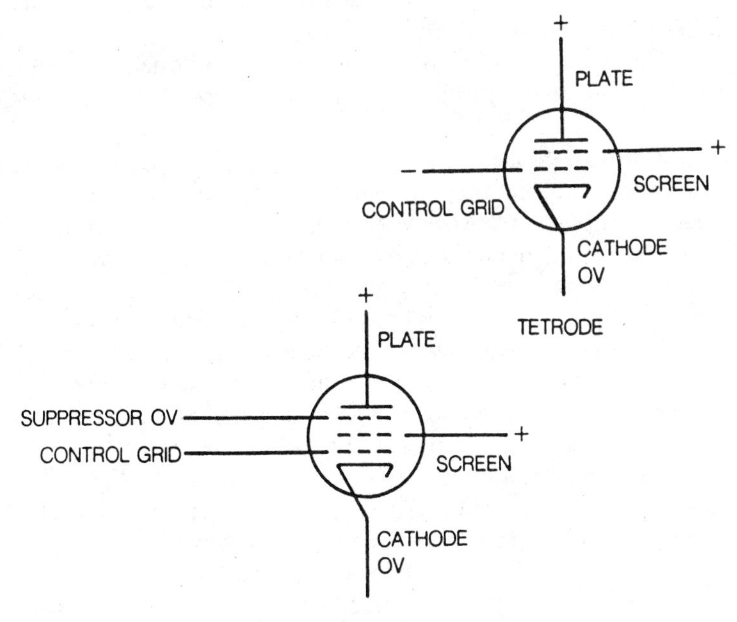

Fig. 2-12. These tubes are now obsolete, but you may have to service tube equipment.

In the tetrode you see that a screen has been added between the control grid and the plate. It is an example of a *Faraday screen.* It prevents capacitive coupling between the grid and the plate by applying a positive voltage to the screen. In addition to preventing capacitive coupling, it helps accelerate electrons toward the plate. Not shown in the illustration is the fact that that screen is at signal ground potential. Therefore, it acts as a grounded electrostatic screen to the signal and a positive screen for electron acceleration.

The pentode shown in Fig. 2-12 has a third suppressor grid added between the screen and the plate. One of the problems in the tetrode is that the electrons become accelerated and strike the plate with enough energy to produce secondary emission. The secondary electrons are often attached to the screen so that over a range of operation, the screen current goes up and the plate current goes down.

To eliminate that possibility, the suppressor grid operates at a potential of zero volts. That makes it highly negative with respect to the plate, and it suppresses the secondary electrons. In other words, it forces them back to the plate.

The pentode is very well suited for high frequency operation because the control grid is effectively isolated from the plate.

Partition Noise. The disadvantage of the pentode and most tetrodes is that they produce noise. Specifically, they produce partition noise. At any instant, the number of electrons that arrive at the plate varies by a small amount. That variation is due to the fact that at any given moment some of the electrons will go to the grids. The actual number varies randomly from instant to instant. Therefore, the plate current varies from instant to instant by a small amount. That small variation is called noise. Partition noise is also a problem with bipolar transistors that will be discussed in the next section.

The screen grid and suppressor grid shown in Fig. 2-12 have nothing to do with the signal. They are simply added to produce better internal operation of the tube itself.

The pentode, tetrode, and triode are all three-terminal devices. The input signal is normally on the control grid. The output signal is normally at the plate. However, regardless of the way the device is used, there are only three active terminals as far as the tube is concerned. They are the cathode, control grid, and plate.

If you should ever have to troubleshoot a system that uses vacuum tubes, remember that the most vulnerable part is the filament. The filament, which heats the cathode, has a limited lifetime. You cannot always tell if the filament has been burned out by touching the tube to determine if it is hot. Some low voltage tubes operate with filament voltages and heat so low that you cannot feel a temperature on the tube envelope. The only way to check the filament is by checking for filament continuity. Remember that the dc voltages on the tube electrodes must be correct if the tube is doing its job. In the following discussion, for all of the devices discussed in this section, the input electrode will be considered to be at zero volts.

The control grid is negative with respect to the input electrode, or cathode, and the plate is positive with respect to the cathode. These are not necessarily the voltages with respect to ground, or common. As far as the circuitry external to the tube is concerned, electrodes can have different voltages. But, the voltages with respect to the cathode are always as shown. In other words, the cathode can always be considered to be zero volts when you are determining the polarities of voltages on the electrode.

The Cathode-Ray Tube. The cathode-ray tube is a form of vacuum tube. It is similar in operation, but not in construction, to the tetrode. Figure 2-13 shows a simplified diagram of the cathode-ray tube and its operating voltage polarities. If you consider the cathode to be zero volts, then the control grid is negative with respect to the cathode. In addition to the control grid, there are accelerating and focus grids.

If you would take the cathode-ray tube (CRT) apart, you would see that these grids are not the wire mesh type that you would find in vacuum tubes used for amplifiers. Instead, they are cylindrical in shape and they have small apertures through which the electron beam passes. This results in beam focusing. The reason focusing is necessary is that the electrons are negative and they repel each other. In order to bring them back to a single small region on the face of the oscilloscope, it is necessary to use voltages on the electrodes to focus the beam.

Deflection circuits move the beam left, right, up, and

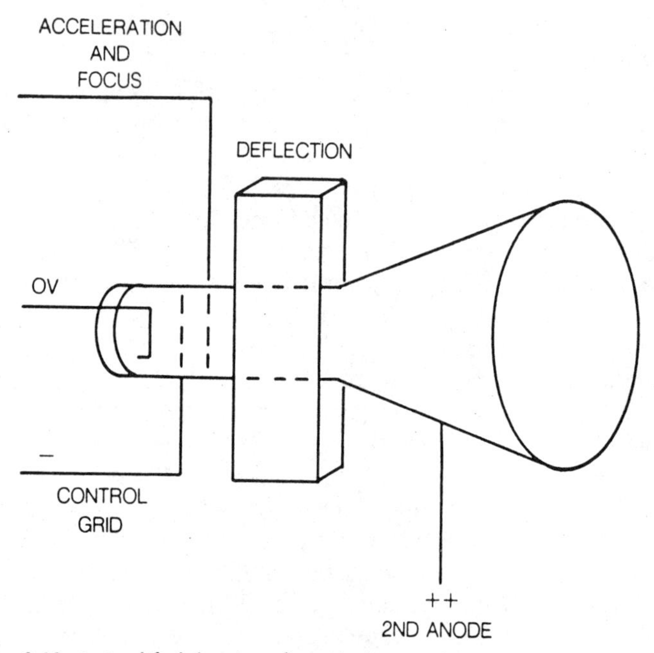

Fig. 2-13. A simplified drawing of a cathode-ray tube.

down. Of special interest is the anode, which is usually referred to as the second anode. It collects secondary electrons that occur when the electron beam strikes the face of the tube. This would be equivalent to the plate of the tetrode. Except for the construction, the operation of the tube depends upon dc voltages just as the tetrode or pentode.

Bipolar Transistors. Bipolar transistors were the first of the solid-state three-terminal amplifying devices. To understand the important difference between this device and the vacuum tube triode, remember that the grid voltage on the vacuum tube controls the grid current. Therefore, it is said to be a voltage-operated device. Bipolar transistors are current-operated devices.

In Fig. 2-14 you see a model and symbol for a bipolar transistor. Refer to the model and note that there is a current between the emitter and the collector. That current comprises about 95 percent of the electrons leaving the emitter and going to the collector.

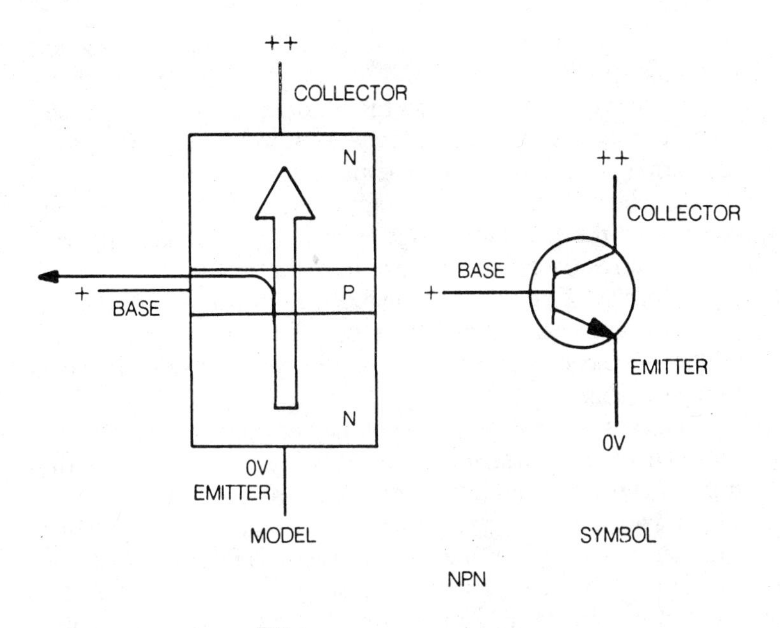

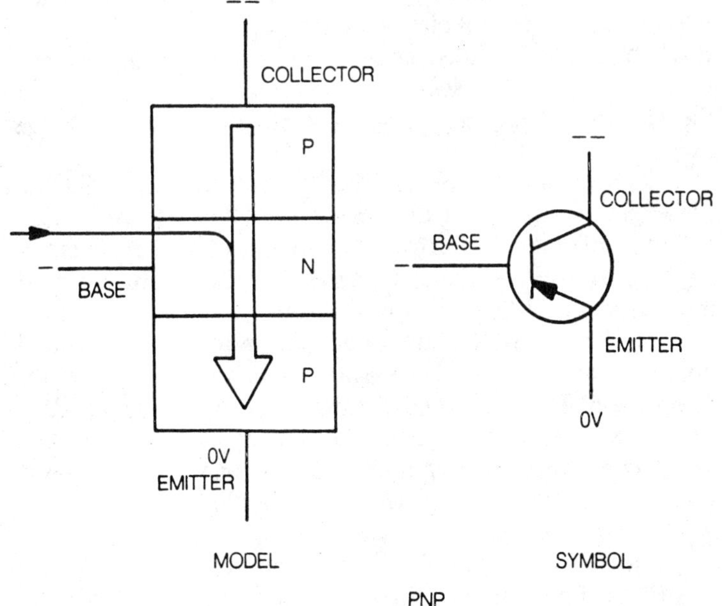

Fig. 2-14. The symbol and model for the operation of a bipolar transistor shows the required operating voltages. Observe the distribution of current represented in the model.

About 5 percent of the electrons go to the base. This is an important point. The number of electrons that go to the collector is directly dependent upon the number that go to the base. Another way of saying this is if you want to increase the collector current you have to increase the base current.

Bipolar transistors are current operated. The key to their use lies in the fact that a small change in base current can produce a relatively large change in collector current.

Observe the polarities of the bipolar transistor. Note that in the NPN transistor, the base is positive with respect to the emitter. The collector is even more positive, as indicated by the two plus signs.

You cannot make a direct comparison between the current operated bipolar transistor and the voltage operated vacuum tube. One reason is an inverse device called the PNP transistor. There is no equivalent vacuum tube device. (Refer to Fig. 2-14.) The electron current in the PNP transistor is from the collector to the emitter. A base current is still needed to control the amount of current that flows through the transistor. Actually, PNP transistors are not often represented as shown in Fig. 2-14. It is better to think of them as being operated with hole currents. In other words, the holes are the majority charge carriers and the electrons depicted in Fig. 2-14 are the minority charge carriers.

Look at the symbols for the two devices. Note that the arrow points away from the base in the NPN transistor. You will remember that the arrow always points toward the N material and away from the P material. The base is made of P material in the NPN transistor. Conversely in the PNP type the base is made with N material and the arrow in the symbol points toward it. Note that negative voltages are required for operating PNP devices. The fact that there are two kinds, PNP and NPN, has led to some very interesting and useful circuits called complementary circuits, in which one of each type is used. You cannot do that with the field effect transistors that will be discussed next.

The JFET. Figure 2-15 shows the models and symbols for JFETs (Junction Field Effect Transistors). These devices are very similar in operation to triode tubes.

Observe the voltages on the source, gate, and drain of the

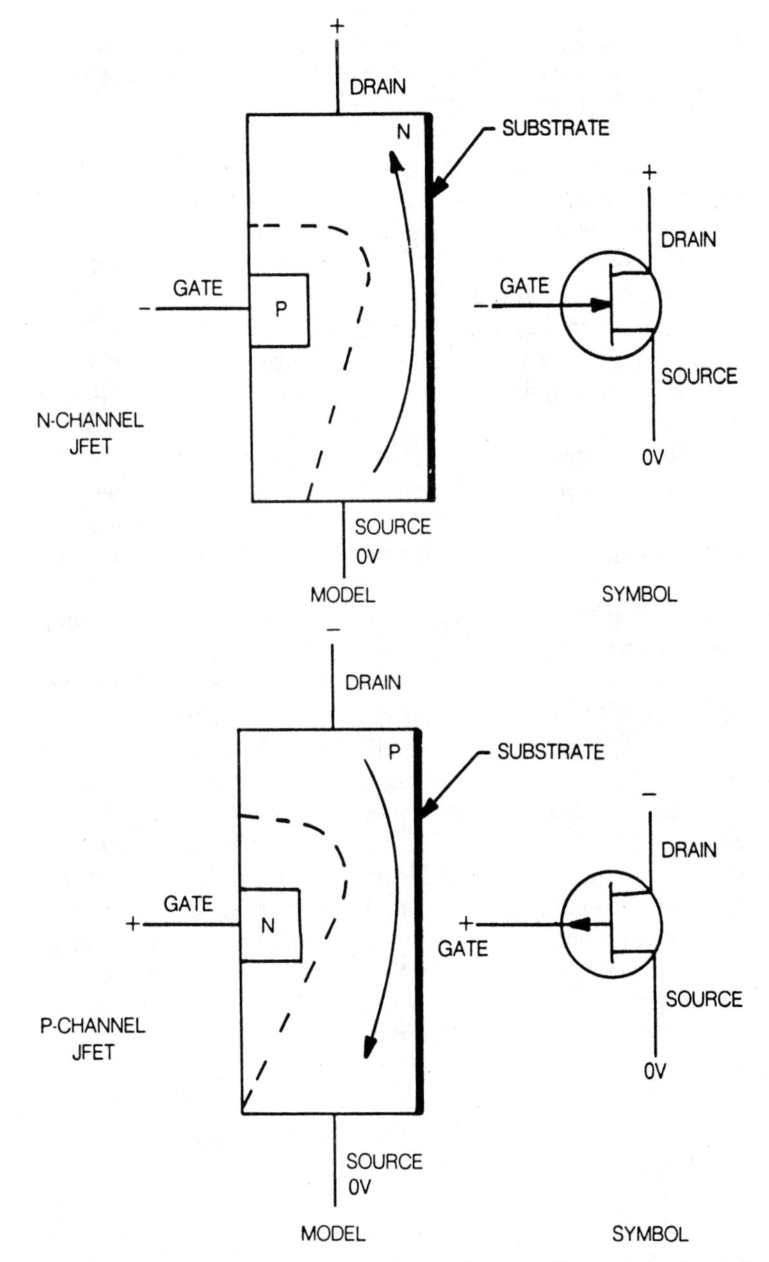

Fig. 2-15. The required operating voltages are shown on this symbol and model for the JFET.

N-channel JFET. Note that the gate is negative with respect to the source and the drain is positive with respect to the source. These are the same polarities that you encountered on vacuum tubes operated as three-terminal devices.

The long arrow on the JFET symbol shows the flow of electrons through the N channel. The shaded area around the gate represents a depletion region. Making the gate more negative increases the size of that depletion region. This, in turn, reduces the space through which the electrons can flow. The gate can be made sufficiently negative to cut off the device. That is another similarity with the vacuum tube. In the JFET symbol, the arrow points toward the channel which is N-type material.

The P material in the P channel JFET is the inverse of the N material. There is no vacuum tube equivalent for this device. Note that the arrow representing electron current points away from the drain and toward the source. The direction of the holes is opposite. Note that the polarity on the P channel JFET requires that the gate be positive with respect to the source and the drain be negative with respect to the source.

Since the gate voltage controls the size of the depletion region, JFETs are voltage operated devices. They are useful because of the fact that a small change in gate voltage will produce a relatively large change in the drain current.

The substrate is a very important part of the operation of all field effect transistors. The name *field effect* indicates that control is by an electric field in the transistor. That field exists between the gate and the substrate. Normally, the substrate is operated at ground or common potential, or at the same potential as the emitter. In the normal operation of the device, you will not have anything to do with the substrate. In most cases it is automatically connected internally to the source of the device.

The MOSFET. A disadvantage of the JFET is the fact that whenever the gate becomes forward biased, a destructive current can flow from the source to the gate. Even if the current is not destructive, it will result in a highly distorted output signal because not all the current leaving the source arrives at the drain.

To get around this problem, an insulating area is fabricated around the gate in a MOSFET (see Fig. 2-16). It prevents accidental gate current flow from peaks of input signals.

Except for that protection, its operation is the same as for the operation of the JFET. The gate voltage controls the size of the depletion region and therefore controls the amount of drain current. This is true for both the N-channel and P-channel devices. The only difference between the two types is in the polarities of the voltages required for their operation.

Of course, the P-type has holes for majority charge carriers and N-type has electrons for majority charge carriers. Note the symbols for the two devices in Fig. 2-16.

The gate part of the symbol is shown insulated from the channel. In fact the original name for these was Insulated Gate Field Effect Transistor (IGFET). The name metal oxide comes from the fact that the gate insulating material is made from an oxidized metal.

The MOSFET discussed in Fig. 2-16 has one disadvantage. In the absence of any gate signal there is only a semiconductor material between the source and the drain. That means that it will conduct current even though there is no signal on the gate. In fact, that current can be destructive in some applications such as power amplifiers.

The term Depletion MOSFET comes from the fact that increasing the gate voltage or bias causes a depletion of the region through which the semiconductor charge carriers can flow. The MOSFETs in Fig. 2-16 are depletion types.

Another type of MOSFET is illustrated in Figure 2-17. It is called the enhancement-type MOSFET. It eliminates the disadvantage just discussed. In this case, the device is made in such a way that the depletion region reaches completely through the channel when there is no gate-to-source voltage. So, in the absence of a gate voltage the device is cut off.

In order to get the device into operation—with current flow through the channel—the gate must be forward biased with respect to the source. For example, in the N-channel enhancement MOSFET, the gate must be made positive in order to reduce the size of the depletion region and allow electrons to pass through. Enhancement MOSFETs are very common in power amplifier circuits. The term *enhancement* MOSFET

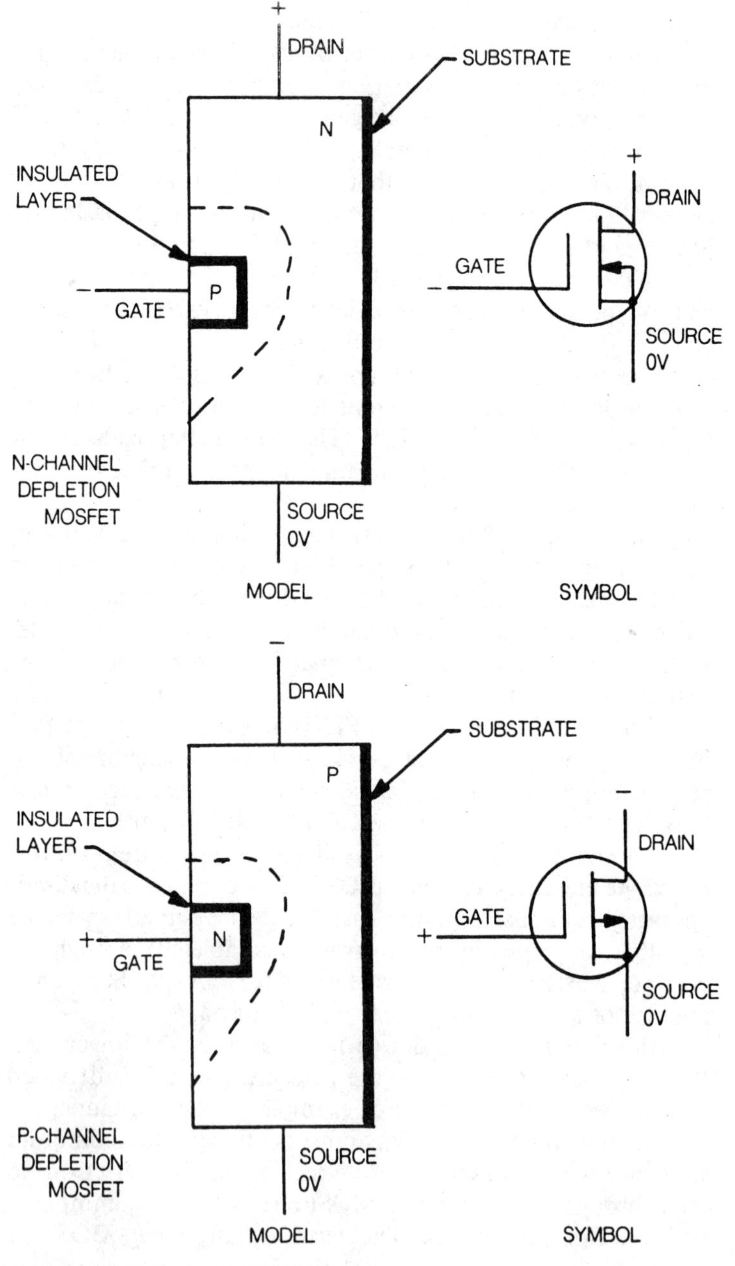

Fig. 2-16. The depletion MOSFET.

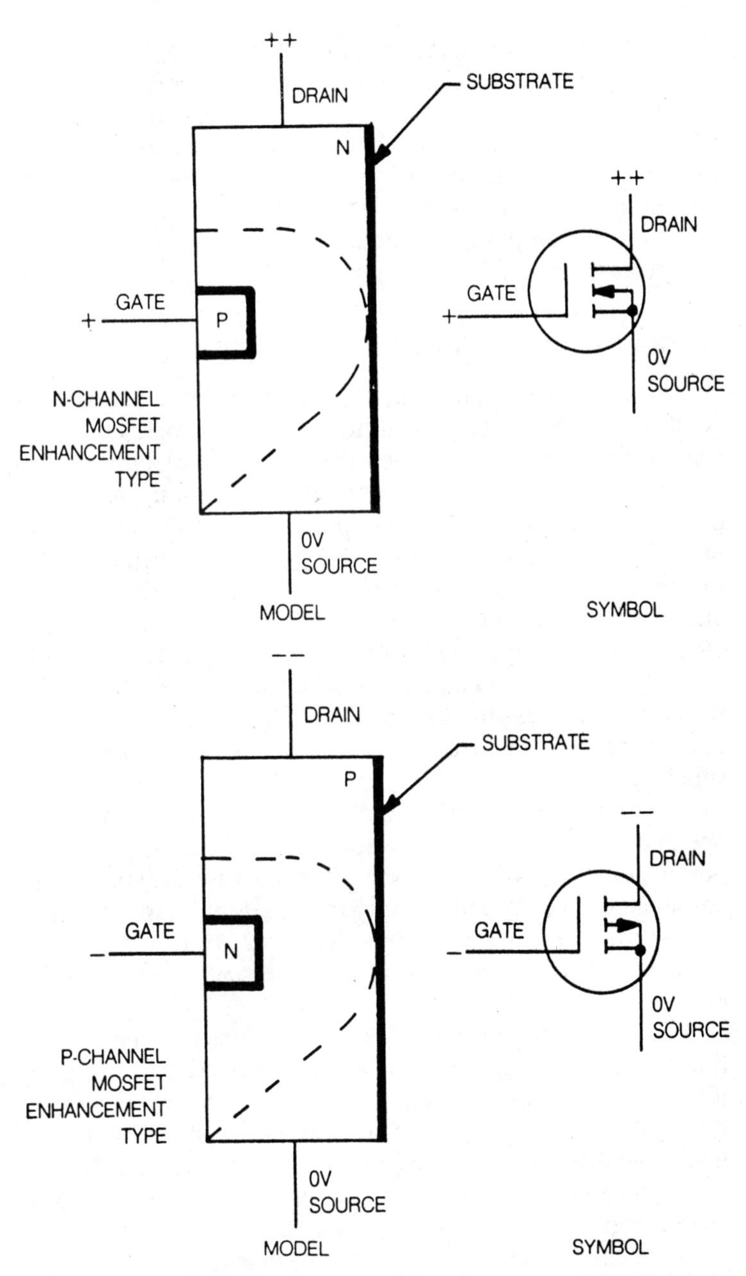

Fig. 2-17. Compare this representation of the enhancement MOSFET with the depletion MOSFET in Fig. 2-16.

comes from the fact that the forward bias increases, or enhances the region through which the charge carriers can flow.

A special type of enhancement MOSFET is called the VFET. Other names are used by different manufacturers. Basically, it is the enhancement MOSFET with a special shape of gate electrode that allows it to dissipate more heat. VFETs are used as power amplifiers in circuits that were originally limited to bipolar transistors and vacuum tubes.

OPERATIONAL AMPLIFIERS

Although operational amplifiers are complete circuits, they are used as components. The term operational amplifier comes from a mathematical use when they first became popular. In mathematics, an *operator* is a symbol that tells you what math procedure to perform. For example, in 6×8, the times sign means multiply. As another example, in the problem $26 - 23$ the minus sign tells you to subtract 23 from the 26. Operational amplifiers can perform all the basic mathematical operations. Hence the name operational amplifier.

Figure 2-18 shows an op amp schematic symbol. Normally the plus and minus power supplies are not shown on the symbol but they are given here to indicate that this device requires both supplies.

In recent years there have been a number of single-ended operational amplifiers that will operate on a positive-only power supply. However, the positive and negative supply is still considered to be an optimum arrangement for most op amps.

Also shown in Fig. 2-18 is the *bode plot* of gain versus frequency. The gain is given in both DB and in voltage gain (A_v).

An important feature of this characteristic curve is the linear rolloff. This makes it an easy matter to estimate the gain if the desired bandwidth is known, or to estimate the bandwidth if the gain is known. A gain of 100 is represented with a broken line. Note that it reflects from the characteristic curve at 10,000 hertz. Therefore, the approximate bandwidth of the op amp is 10,000 hertz when it is has a gain of 100.

The bandwidth indicated in Fig. 2-18 is not precisely true. You know that the bandwidth of an amplifier is the range of

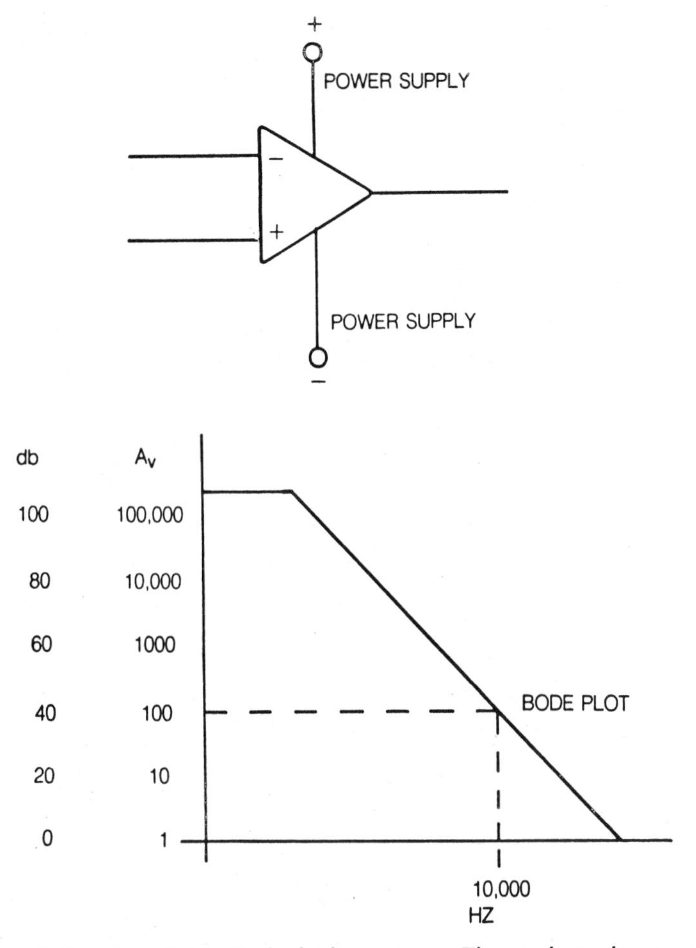

Fig. 2-18. The symbol and bode plot for an op amp. The parady supply connections are not normally shown.

frequencies between the points where its voltage characteristic curve drops to 0.707 times its maximum value. So, in this case the bandwidth would be slightly wider than shown in the illustration.

At the low end of the frequency range, the amplifier response curve does not drop off. That is because inside the op amp there are direct-coupled amplifiers from the input to the output. For op amps then, the bandwidth starts at 0 hertz.

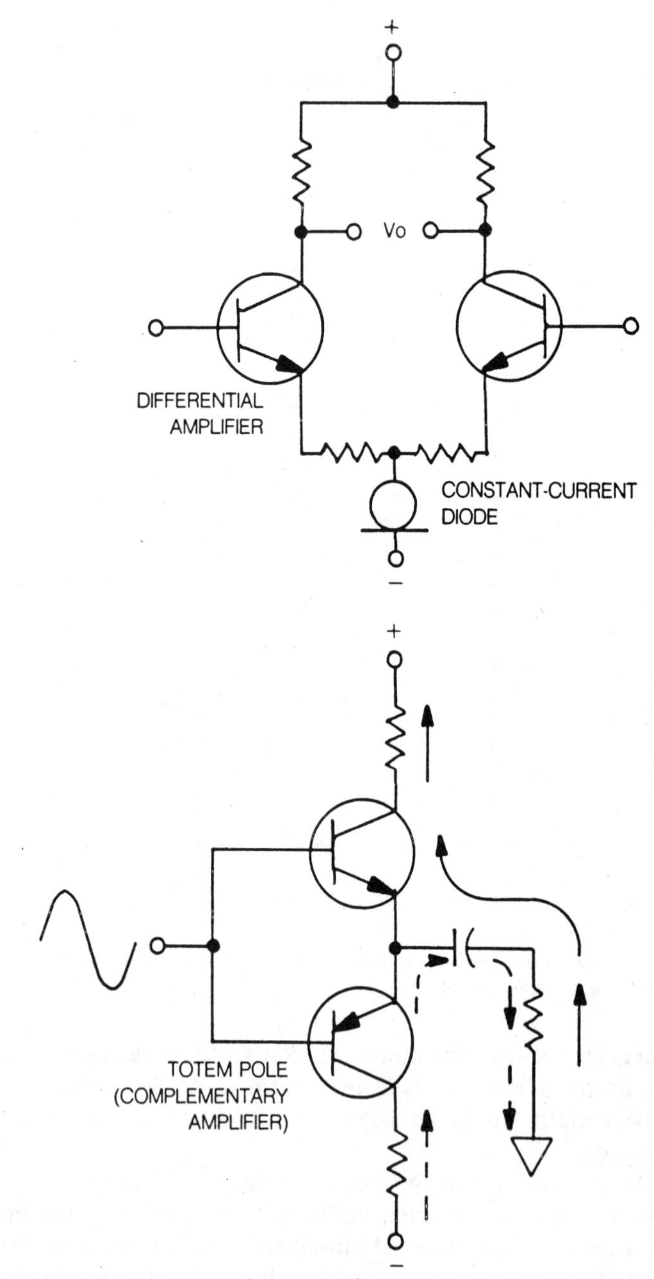

Fig. 2-19. Two circuits used in a typical op amp.

There are two important circuits in an op amp that should be considered before discussing their use. They are illustrated in Fig. 2-19.

The differential amplifier is well known to technicians. The ones used in operational amplifiers employ "long tail" (plus and minus) power supplies. In this example, a constant-current diode is used to supply a fixed amount of current to the junction of the two emitter resistors. This is important. When an unbalancing signal is applied to the differential amplifier, the amount of current increase in one of the transistors is exactly equal to the amount of decrease in the other transistor. The total current is a fixed value.

One way to use this differential amplifier is to ground one of the bases and apply the signal to the other. You can ground the bases without cutting the transistors off because the base voltages are half way between the positive and negative supply voltages. In a properly designed differential amplifier the two base voltages are at approximately zero volts.

The output voltage (V_o) is proportional to the difference in the two signals at the bases. Inside an op amp, this output voltage is delivered to direct-coupled amplifiers and then to the output. A totem pole circuit, also shown in Fig. 2-19, is used for the op amp power output.

In the totem pole circuit, two transistors are stacked in a complimentary configuration. Note, again, the long tail bias. Therefore, the inputs for the bases are very close to zero volts and the output is at zero volts. For this reason you can take a screwdriver and short circuit the output to ground and there is no damage to the power amplifier configuration.

Assume that the input to the power amplifier is a sine wave with positive and negative half-cycles. On one half-cycle, the signal goes positive. This causes the upper transistor to conduct and the lower one to be cut off. The conduction path charges the capacitor in an electron path shown by the solid arrow.

On the next half-cycle the input signal goes negative, cutting off the upper transistor and causing the lower transistor to conduct. This discharges the capacitor in an electron path shown by the broken arrow.

Obviously, the capacitor must be a very large value in order to be able to hold the full charge and discharge. In many applications, this capacitor is an electrolytic type. The circuits in the two figures of 2-19 are simplified. In reality they are more complex than shown but the principle of operation is the same.

Typical Circuits. Figure 2-20 shows some typical op amp connections. The common-mode connection is used to test the ability of the op amp to reject a common-mode signal. If the op amp is perfectly balanced and the two inputs are connected together as shown in this illustration, it is called a common-mode circuit. The output should be zero volts.

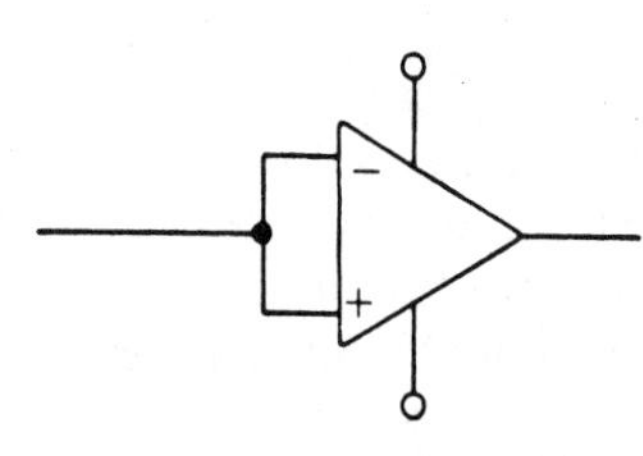

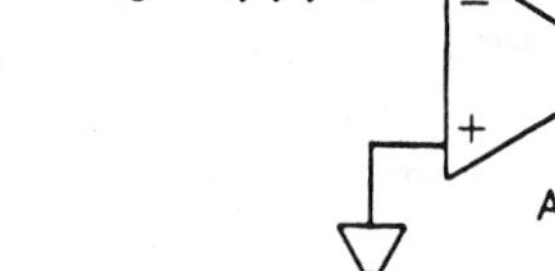

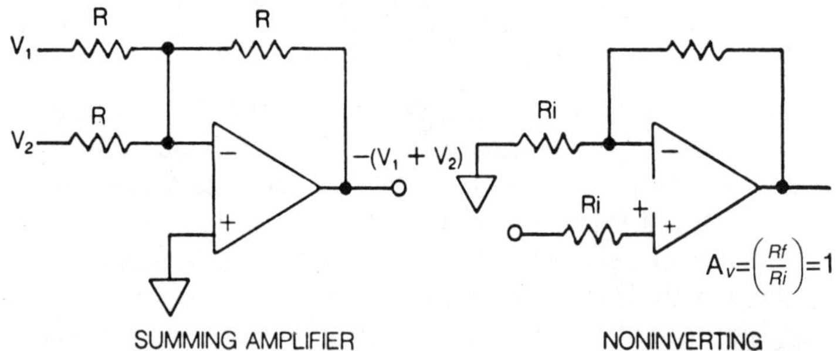

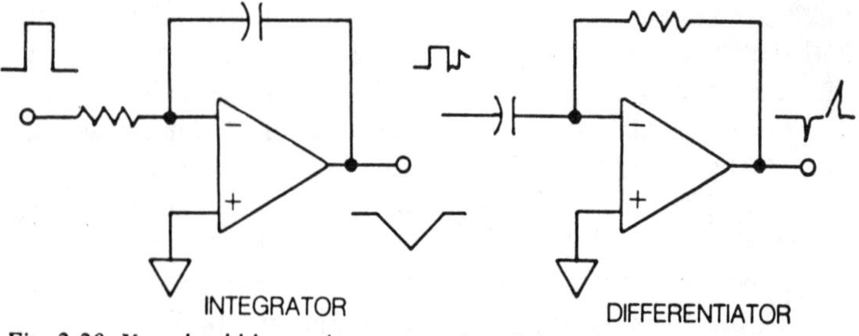

Fig. 2-20. You should know these connections for op-amp circuitry.

In a typical common-mode test, a low-amplitude ac signal is applied to the common-mode input and the output is measured. The ability of the op amp to reject the common-mode signal is a desirable feature, called the common-mode rejection ratio.

For the inverting amplifier, the input signal is to the terminal marked with the negative sign. With this connection, the output will be 180 degrees out of phase with the input signal. Note the simple calculation of voltage gain with this op amp. It is only necessary to divide the feedback resistance by the input resistance in order to get the voltage gain. The negative sign in the equation means that there is a 180 degree phase shift.

This simple equation would not be possible except for the very high open-loop gain of the op amp. Basic theory of amplifier design says that whenever the amplifier gain is very high, the amplifier characteristic is completely dependent upon the feedback circuit. The high open-loop gain was shown in the characteristic curve of Fig. 2-18. Much higher gains are possible with modern operational amplifiers. The one shown is characteristic of the 741, which has become the standard for the op amp industry. In early vacuum tube equivalent op amp circuits, gains of 5000 were considered to be adequate.

The summing amplifier of Fig. 2-20 demonstrates its ability to perform a basic operation in arithmetic. The output is directly related to the sum of the inputs. The negative sign indicates phase inversion.

Op amps can be connected as noninverting amplifiers. The connection is shown in Fig. 2-20. The gain of the noninverting amplifier is given on the drawing. It is very slightly greater than the gain of the inverting amplifier. As an industrial electronics technician, you are expected to know how to calculate the gain of op amps and how to determine their bandwidth from a bode plot.

Two very important op amp applications are the differentiator and integrator. (See. Figure 2-20.) The differentiator is used extensively for converting a pulse- or square-wave input to positive and negative trigger voltages. Normally, the negative trigger is eliminated with a diode, so the circuit produces a positive trigger voltage that is relative to the frequency of the incoming pulse.

It is also possible to use the differentiator in a number of other applications. For example it can be used as a high-pass filter. It produces an output voltage, proportional to the rate of change of input voltage. So, differentiators can be used to find the derivative of an input waveform that represents some changing quantity.

Differentiating amplifiers are sometimes used in robotics because of the fact that they produce a rapid change in output voltage for low voltage values, but as the value increases, the rate of change decreases. In a robot application, this might be a desirable feature if used for moving a robotic arm. When the arm first starts to move it can be moved rapidly. Therefore, the total time necessary for moving the arm from one point to the other is reduced. As the arm approaches its destination it is desirable to slow its motion in preparation to locking onto its final destination.

The term integrate means to summarize. The integrator produces an output signal, related to the sum of all of the input signals, or to the sum of all of the parts of the input signal. An example of its use is in converting square waves to triangular waves, and other waveform conversions.

This has been a short review of op amps. If you are not familiar with these components, you should take the time to give them some in-depth study. As an industrial electronic technician you are expected to know how they work, how to calculate their gain and the distinguishing features of the basic circuits shown in Fig. 2-20. This is by no means the extent of their possible uses. In fact, one of the most important things about the op amp is that it is so versatile that it can be used in an almost unlimited number and variety of applications.

Basic Troubleshooting Ideas. When you are troubleshooting industrial electronic equipment that has three-terminal linear components, remember that there are two voltages present in their normal operation. One is the dc operating voltage, such as the bias and the output electrode voltage. This voltage must be correct or the device cannot operate. The second voltage present in these devices is the signal voltage.

You can easily trace the signal with an oscilloscope. Compare the input and output signals throughout the system. If the system is not operating properly, and the fault is one of the three-terminal linear devices discussed in this chapter, then you should be able to find it by measuring the dc voltages and tracing the signal path.

SUMMARY

In most applications, the vacuum tube has been replaced by solid state devices. However, there are still some systems that use tubes. You should know a little about how they work. Their operation has been briefly reviewed in this chapter.

Magnetrons are vacuum tube diodes. Cathode-ray tubes have the same elements as a triode, but they have a screen instead of a plate. There are still many transmitters in operation that use tubes. You might run across them in an industrial two-way radio system. There is still a lot of expensive equipment in industry that has tubes. The manufacturer isn't going to replace that equipment just because there is a new solid state version.

For each of the three-terminal devices, the normal operating voltages are included with their symbol. Those are the operating voltages you would find when the device is used as an amplifier. In other applications, the voltages can be different. For example, the base-collector junction of a bipolar transistor might be forward biased in switching circuits. In amplifiers that junction is reverse biased.

There has been some discussion of diodes in this lesson. More diodes are included—especially in the chapters on thyristors and power supplies.

SELF TEST

1. A magnetron is a type of:
   (A) semiconductor diode.
   (B) vacuum tube diode.
   (C) triode.
   (D) JFET.

2. The method of obtaining a high current flow by electron-atom or electron-molecule collisions is called:
   (A) regeneration.
   (B) degeneration.
   (C) reformatting.
   (D) avalanching.

3. Although the mechanism for operating is different, a triode tube does the same job as a:
   (A) bipolar transistor.
   (B) JFET.
   (C) depletion MOSFET.
   (D) All of the choices are correct.

4. The current through a PNP transistor is mainly composed of:
   (A) electron flow.
   (B) hole flow.
   (C) negative ion flow.
   (D) No choice is correct.

5. Which of the following is correct?
   (A) A bipolar transistor is cut off when there is no base bias.
   (B) A JFET is cut off when there is no gate bias.
   (C) Both choices are correct.
   (D) Neither choice is correct.

6. Is the following statement correct? The operating voltages for an N-channel JFET are the same as for an N-channel depletion MOSFET.
   (A) Correct.
   (B) Not correct.

7. In the normal operation of an NPN transistor (as an amplifier) the base-collector junction is:
   (A) not biased.
   (B) forward biased.
   (C) reverse biased.

8. Which of the following is NOT forward biased during normal operation?
   (A) enhancement P-channel MOSFET.
   (B) PNP transistor.
   (C) depletion P-channel MOSFET.
   (D) All must be forward biased.

9. Cathode ray tubes get electrons for current flow by using:
   (A) field emission.
   (B) photoemission.
   (C) avalanching.
   (D) No choice is correct.

10. The − and + inputs on an op amp represent:
    (A) power supply voltages.
    (B) polarities of input signals.
    (C) inverting or noninverting operation.
    (D) None of these choices is correct.

ANSWERS TO SELF TEST

1. (B) Magnetrons are not conventional vacuum tube diodes but they have a cathode and anode in a vacuum chamber. By definition, that is a vacuum tube diode.

2. (D) Avalanching occurs in semiconductor diodes as well as in gas-filled diodes. In fact, there is a family of semiconductor devices called avalanche diodes and avalanche transistors. They are not popular in modern systems.

3. (D) The JFET is very similar to the vacuum tube triode. Refer back to their diagrams and note that they have the same voltages on their electrodes. In a triode there will be grid current flow if the grid is made positive with respect to the cathode. Likewise, gate current will flow in an N-channel JFET if its gate is positive with respect to its source.

4. (B) Hole flow is a very important model for current flow in P-type semiconductor materials.

5. (A) Current will flow through the channel of a JFET when there is no bias. The channel is made of N-type or P-type material.

6. (A) You should memorize all of the operating voltages for the two- and three-terminal devices discussed in this chapter.

7. (C) In switching circuits the base-collector junction may be forward biased.

8. (C) Be sure you know the difference between depletion and enhancement MOSFET types. That includes differences in biases and symbols.

9. (D) Cathode ray tubes have cathode emitters. Electrons are obtained by thermionic emission.

10. (C) When the input is at the inverting terminal, the output signal is 180 degrees out of phase with the input. If the noninverting input is used, the output and input signals are in phase.

3

Thyristors and Other Switching Components

SWITCHING IS A very important part of industrial electronics circuitry. It is performed by many types of devices that are the subjects of this chapter.

In their simplest form, switches are mechanical On/Off devices that permit us to control the distribution of electrical energy. But not all switches are simple On/Off devices. Many switches used in industrial electronics simply switch from one voltage level to another. Furthermore, not all switches are manually operated. Many of the switches change from one position to another by the application of a voltage or a pulse. Relays have always been an important part of switching circuitry in industrial electronics. In many circuits they have been replaced by semiconductor devices. There are some characteristics of relays that make them still useful today.

You can think of a relay as being a current amplifier. A small input current, used for switching the relay On and Off, can be used to control a very large output current that switches a power device. For example, a relay may be activated by a few hundred milliamps and control a 15 ampere light circuit.

Relays can also be thought of as interfaces. The input voltage may come from a semiconductor device such as a logic circuit that cannot supply high current. Its output, though, can be used to control motors, lights and other high-current devices.

Many relays are made with large coils and heavy contacts, but modern relays are also made in very small packages. Examples are reed relays and relays in integrated circuit packages.

The electronic switches of industry are called thyristors. They are also called breakover devices. Thyristors will not conduct until the applied voltage reaches a certain minimum predetermined value. Once that value is reached they conduct heavily.

The voltage for conduction may be predetermined at the manufacturing of the device. This is the way it is done in two-terminal breakover devices such as diacs and four-layer diodes. On the other hand, the breakover point may be established by the presentation of a pulse which changes the device from an Off to an On condition. SCRs and SCSs are examples.

You may be surprised to find a discussion of neon lamps and lasers in this chapter. The earliest lasers and many of the lasers made today are pulsing devices. They do not produce a continuous output. For that reason alone, lasers are included here.

Also, lasers operate by switching the energy levels of electrons. That is another reason for including them in this chapter.

CHAPTER OBJECTIVES

After you have studied this chapter you will know:

- How a laser works and the relationship between a laser and a neon lamp.
- The difference between an SCR and an SCS.
- Why relays are never used in integrated circuit logic systems.
- The difference between 3-layer and a 4-layer diodes.
- The differences and the similarities between a UJT and a PUT.

SWITCHES

From our experience with switches in daily life such as the headlight switch on a car and the light switch in the house, we have come to think of a switch as a simple On/Off device. Actually, there are many switches in electronics that switch between two voltage values but are never in a full On or full Off condition.

One example is the tunnel diode that was discussed in chapter 2. It is one of the fastest switching devices that can be purchased off the shelf.

Mechanical switches are not suitable for use in integrated logic circuits. The simple reason is that contacts come together and then bounce in those switches. As will be discussed in a later chapter, there are many switches available in logic circuitry that are "bounceless." However, if you connect a mechanical switch into a logic circuit, the bounces will operate the logic system and cause unpredictable results.

Reed Switches. Reed switches are made with two thin strips of metal enclosed in a glass envelope. These switches are small and come in many different sizes.

The operation of a reed switch as a proximity switch is illustrated in Fig. 3-1. In this application, the contacts of the

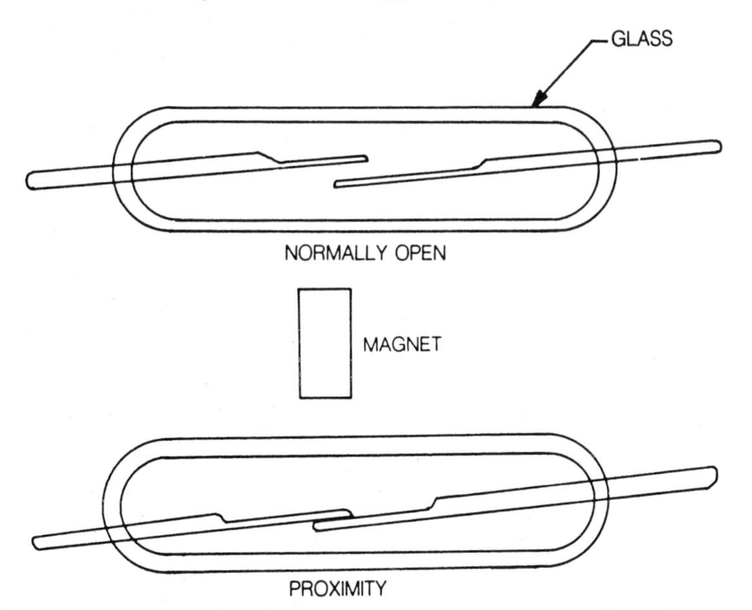

Fig. 3-1. A reed switch with normally open contacts is shown in the top illustration. As shown in the bottom illustration, when a magnet is near the reed switch, its contacts close.

switch are normally open. When a magnet is brought close to the switch, its magnetic field induces a magnet field in one of the reed contacts. That causes it to bend and touch the other contact. This is also illustrated in Fig. 3-1.

These simple proximity switches can be used to limit the travel of a machine, to sense when a door is open or closed, and many other applications in industrial electronics.

Another way of using the reed switch is illustrated in Fig. 3-2. Wrapping a coil around the switch gives it a configuration called a reed relay. In this case, a current through the coil, from a to b, produces a magnetic field that causes the contacts to close. Reed relays are very sensitive: it does not take a large amount of current to close the switch contacts.

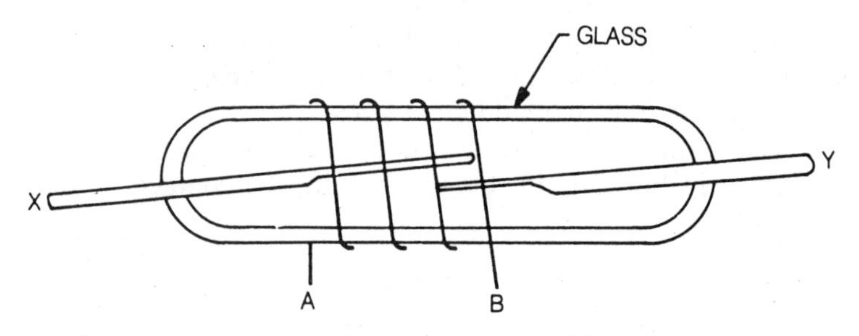

Fig. 3-2. A current-carrying coil provides a magnetic field to close the reed switch terminals. This is a simple form of relay.

Thermal Switches. Whenever two dissimilar metals are clamped together to produce a bimetal strip, they will bend when heated. This bending action occurs because the coefficient of expansion for one of the metals is greater than for the other.

This characteristic of bimetal strips can be used in making a switch. Figure 3-3 shows an example. The switch contacts are closed when the device is cold. When it is heated, the bimetal strip bends and opens the contacts. A simple switch like this can be used to sense overheating in an electronic system. It can also be used as an alarm when placed in an area where there is supposed to be heat, but the heat is, for some reason, removed.

Note the contact symbol in this switch. The switch contacts look very much like a capacitor. However, the correct symbol for a capacitor has one curved and one straight plate. Despite the popularity of this switch contact symbol, some foreign man-

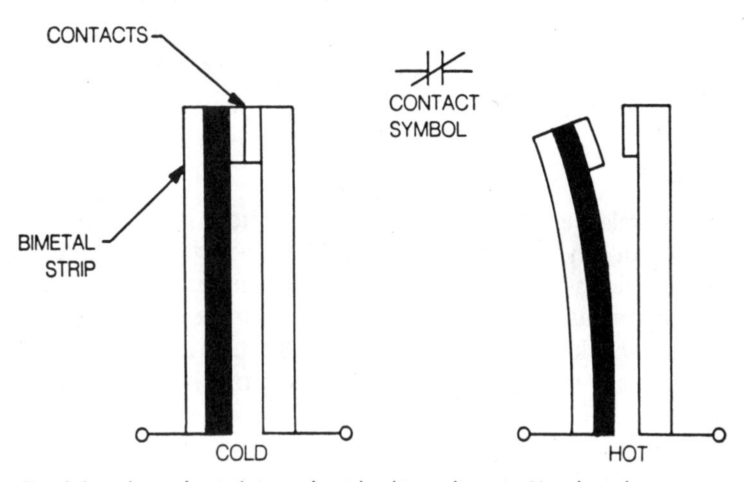

Fig. 3-3. A thermal switch is made with a bimetalic strip. Note how the contacts open when the strip is heated.

ufacturers provide schematics for their equipment that have straight lines for their capacitors. As a technician you have to be aware of the fact that some symbols used by one manufacturer are not universal for all other manufacturers. It is your job to analyze the schematic symbolism as well as the circuitry.

Figure 3-4 shows how the thermal switch of Fig. 3-3 can be made into a thermal relay. In this case, the leads to a heater are brought out externally to the device. When a current is supplied to the heater, the switch contacts open as shown in Fig. 3-3.

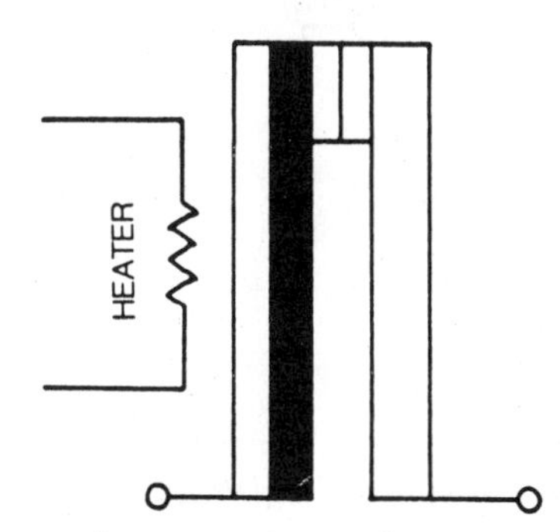

Fig. 3-4. A heater element can be used to operate the thermal switch. When current flows through the heater, the resulting rise in temperature opens the contacts.

This type of relay is slow to operate because it requires time for the heat to affect the bimetal strip. However, it is a reliable switch for some applications.

RELAYS

A relay is an electromechanical switch. Its principle of operation can be seen in Fig. 3-5. When the switch is closed, a current flows through the coil and produces a magnetic field. The coil is usually wound on a soft iron material. The strong magnetic field attracts the armature and causes it to move from its normally-closed position (NC) to its normally-open position (NO).

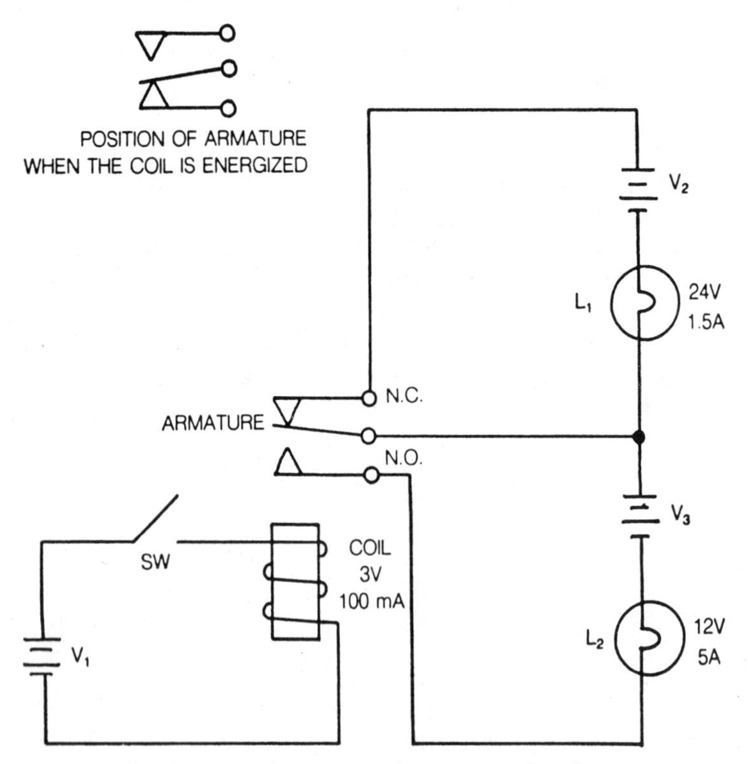

Fig. 3-5. This illustration demonstrates the operation of a relay.

You should remember this very important thing: relays and switches are normally shown in their de-energized or open condition. You have to make the conversion mentally when you are imagining what happens in a circuit when it is energized or closed.

In the circuit of Fig. 3-5, when the relay is energized, lamp L1 will go off. At the same time, lamp L2 will go on because that circuit is completed.

Different power suppliers are shown for the two circuits operated by the relay contacts. However, a single power supply can sometimes be used for both the coil and the circuitry being switched.

An advantage of the relay is an extremely high current gain. In the circuit of Fig. 3-5, only 100 milliamps are required to operate the coil, but it is able to switch circuits that have 1-½ to 5 ampere loads. The switch for the coil can be located in a remote position, making it possible to control current and voltages at a distance.

Relays are not always large and bulky—they are available in integrated circuit packages. The ICs mount easily on printed circuit boards with automated equipment.

Relays are relatively inexpensive and easy to use. A question that is sometimes asked is, "Are they reliable enough for modern electronic circuitry?" Relays today may be rated for millions of operations. That makes them suitable for use in high-reliability electronic systems.

Another very good advantage of the relay is the fact that when it is off, it is really off. Using a relay or switch contact, the off condition produces an enormous amount of impedance and reduces the circuit current to zero.

A disadvantage of the relay is that when it is operated, the contacts bounce. In the simple circuit of Fig. 3-5, that bounce would not produce any serious problems. However, when relays are used in some types of circuits, the contact bounce is highly undesirable.

The problem of contact bounce has been partially eliminated by using mercury-wetted contacts. That way when the armature moves it "splashes" into the mercury and the semiliquid mercury retains contact through the bounce time period.

Fan-In and Fan-Out. An important feature of relays is their high fan-out capability. Fan-out refers to the number of circuits that can be operated simultaneously with a single coil. To accomplish that, many armatures are stacked with corresponding normally-open and normally-closed switch contacts. When the relay is energized or de-energized, all the contacts are operated simultaneously.

Electronic Relays. In modern industrial electronic systems, you will sometimes see devices referred to as electronic relays. They perform the same task as shown in Fig. 3-5 but they do it with electronic components and no moving parts. Examples of electronic relays will be given in the discussion of SCRs.

Ladder Diagrams. Schematics like the one in Fig. 3-5 are not often used as an industrial electronics drawing. Instead, the relays are shown with ladder diagrams. Figure 3-6 shows the symbols that are used. Both types of symbols in Fig. 3-6 are used in industrial diagrams. Ladder diagrams are preferred. They are simpler to draw and show more information in a given amount of space.

Figure 3-7 shows the same circuit as Fig. 3-5 in a ladder diagram form. An important difference in this illustration is the use of a single 24 volt supply. In order to get the coil and lamps to operate across a 24 volt supply, it is necessary to add series resistors. They are R1 and R2 in the diagram. Calculation of these resistance values is a simple Ohm's law problem.

The letters beside the switch contact in the ladder diagram reference the coil that operates those contacts. This is necessary because in ladder diagrams there are often several relays. Each relay has a set of contacts operated by a separate coil. The contacts may be distributed throughout the schematic diagram, so it is necessary to indicate which contacts are operated by which coils.

Ladder diagrams are used for ac and dc circuits. AC relays are different from dc relays. They have laminated cores to reduce eddy current losses, which do not occur in the dc coil.

Maintaining Relays. In older relays it is necessary to burnish the points occasionally—especially if they are switching high currents. Arcing of high currents tends to weld contacts together. If not totally welded together, small bumps are pro-

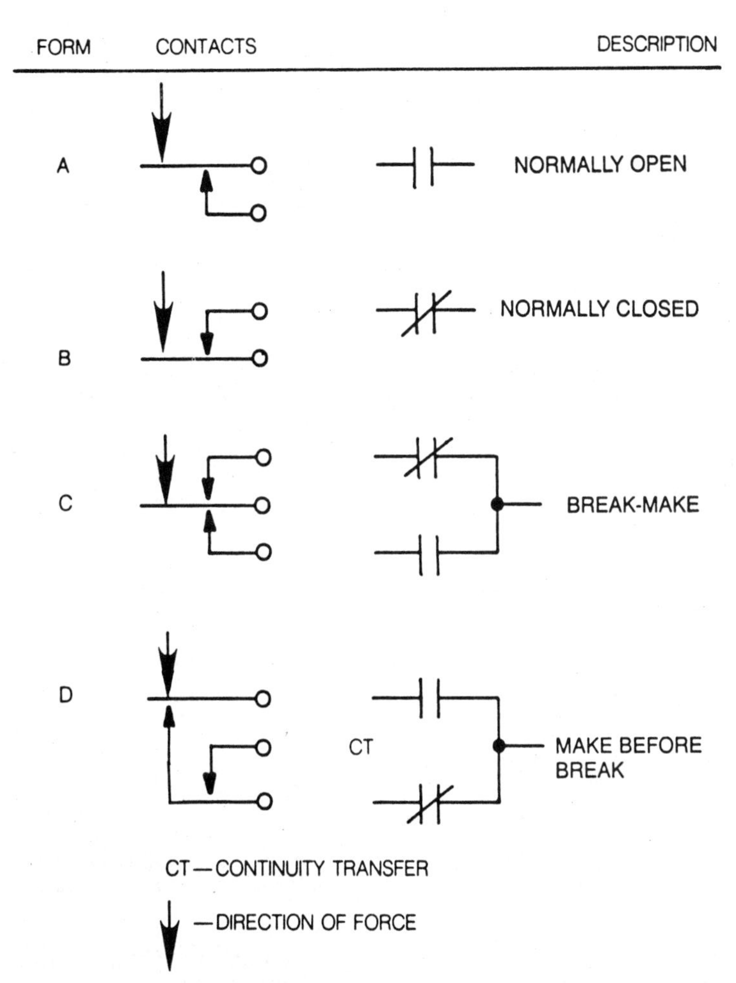

Fig. 3-6. Two methods of representing contacts are shown here. Note the lettered identifications.

duced that must be removed in order to bet reliable contact. You have to burnish (smooth) the contacts by rubbing them with a non-metalic abrasive.

LASERS

Lasers are very easy to understand if you know some basic concepts of atomic physics; specifically, how electrons are stim-

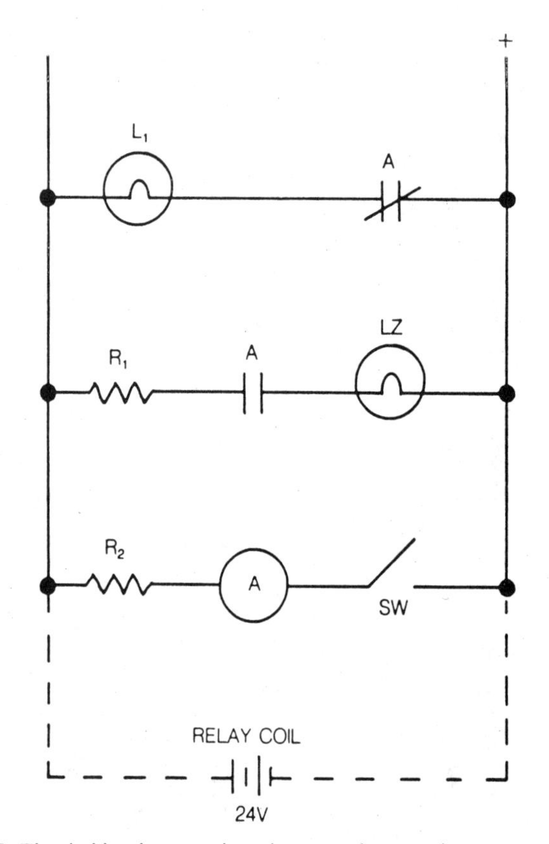

Fig. 3-7. This ladder diagram does the same thing as the circuit in Fig. 3-5. Resistors R1 and R2 were added so that a single power supply could be used.

ulated to produce light. The term *laser* stands for *l*ight *a*mplification by *s*timulated *e*mission of *r*adiation. I will use the neon lamp to explain how the light is obtained.

The Neon Lamp. Neon lamps have been used for voltage regulators, breakover devices, switches, and a host of other applications. However, today, semiconductor devices have largely replaced them in circuitry so that now they are most often used as indicators. The theory of the neon lamp is very important to understanding some semiconductor devices. Also, you may see them in some old equipment, still being used in their original applications.

Figure 3-8 shows a neon lamp in the form used in elec-

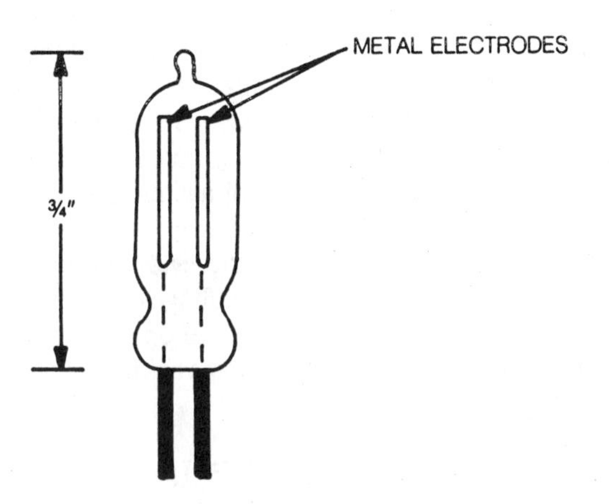

Fig. 3-8. Neon lamps like this found many uses in earlier industrial electronic circuits.

tronics. It is made with two metal electrodes inside a glass envelope filled with neon gas. Two leads for the metal electrodes are brought outside for connection into a circuit.

Figure 3-9 shows the simple circuit that will be used for discussing the characteristics of this device. It is made with a constant-voltage source (V), a variable resistor, (R) and a neon lamp. Observe the symbol for the neon lamp. The dot inside a circle means that there is a gas inside the envelope.

The circuit of Fig. 3-9 makes the current variable. In other words, the current is the independent variable in the circuit and the voltage across the neon lamp is the dependent variable.

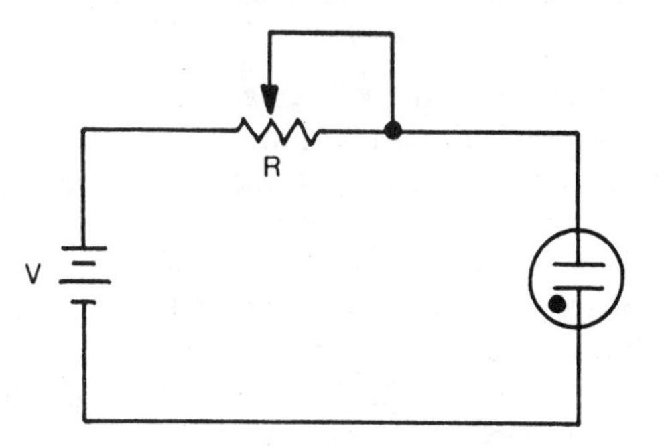

Fig. 3-9. With this circuit, current is the independent variable.

In scientific characteristic curves, the independent variable is traditionally plotted on the horizontal axis and the dependent variable is plotted on the vertical axis. This explains the axis for the graph in Fig. 3-10. The current is increased as you move from left to right, and the resulting voltage drop is increased as you move from bottom to top.

In order to obtain this graph, it is necessary that the applied voltage be greater than the breakover voltage of the neon lamp. This voltage is indicated by a broken line on the graph of Fig. 3-10. For most of these lamps, the breakover voltage is about 70 volts.

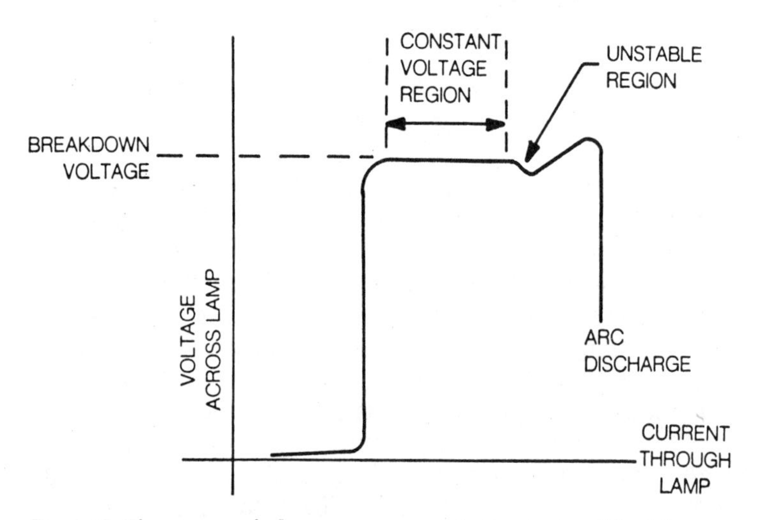

Fig. 3-10. The curve made from measurements in the circuit of Fig. 3-9.

For very low currents, the voltage across the lamp is low. However, at some point the current produces a steep rise in the voltage across the lamp. This occurs due to an "avalanching" condition inside the lamp. Free electrons in the gas are attracted by positive voltage on one electrode. As the electrons move toward that voltage, they strike atoms and knock other electrons loose. Those, in turn, strike other atoms and knock even more electrons loose. Avalanching was illustrated in chapter 2 (Fig. 2-6) and is repeated here in Fig. 3-11.

As you can see from the graph, this immediately produces a constant voltage across the lamp. If you were looking at the

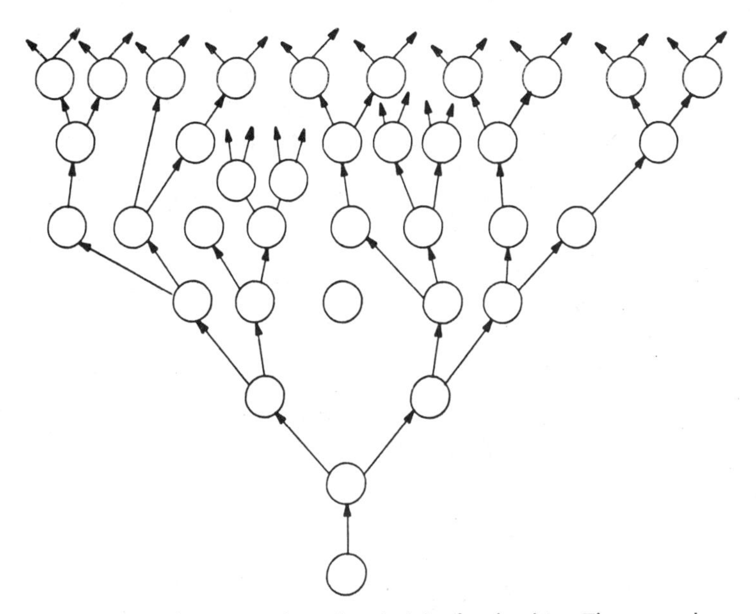

Fig. 3-11. This illustrations shows the principle of avalanching. The arrows show electron paths.

lamp you would see that it lights at this point, so this voltage is sometimes called the glow potential.

A further increase in current produces no noticeable increase in the voltage across the neon lamp over a range of current values. The voltage is constant up to a point where a slight decrease in the voltage occurs. This point is called the unstable region. Lamps are never operated in this region.

Increasing the current beyond this level produces a slight increase in voltage until an arc discharge point is reached. The arc discharge is a tiny lightning bolt inside the neon lamp. Once it occurs, the lamp is no longer usable.

Of special interest to us is the cause of the brightness at the negative electrode. If you look down from the top of the lamp you will see that there is a dark space around the negative electrode. The glow occurs beyond that dark space. The dark space is technically known as Crookes dark space.

Remember that during avalanching, electrons strike atoms and knock other electrons loose. When an atom has lost an electron it becomes a positively charged atom.

The positively charged atom will begin to drift toward the negative electrode. Eventually it reaches a point near the surface of the electrode. Because unlike charges attract, the atom's positive charge causes an electron to jump from the surface of the negative electrode to it. The electron leaving the negative electrode is used to neutralize the positive charge on the atom. This is illustrated in Fig. 3-12.

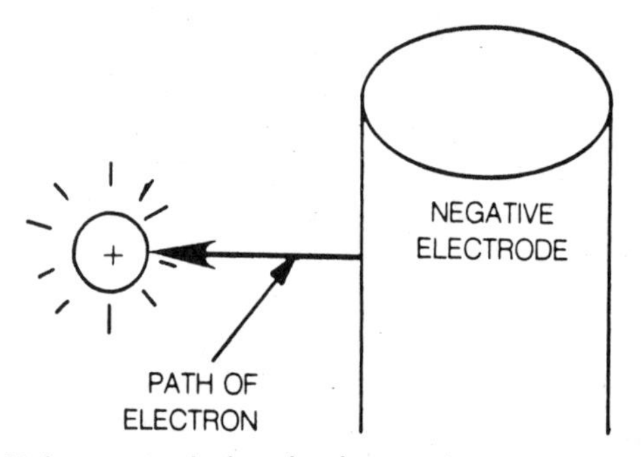

Fig. 3-12. Light is emitted when the electron gives up energy.

The mass of an electron is negligible. So, after leaving the surface of the electrode, its acceleration is very high. Due to its motion, it achieves a high energy. Remember that energy due to motion is called kinetic energy.

When the electron combines with the atom it must give up this high kinetic energy because energy cannot be created or destroyed. The energy is given up in the form of light. This is what produces the light surrounding the negative electrode. It is very important that you understand that the light comes as a result of the electron giving up its kinetic energy when it combines with the atom. This point will be discussed further in connection with another device.

How the Laser Works. There are so many applications of lasers in electronics and in industrial electronics that listing them would require a book to itself. It is very important for industrial electronics technicians to understand how these devices work and a little about how they are made.

The light from an LED consists of individual wavelengths generated at random times. Since the waves are not in phase, they contain very little energy. Out of phase light waves are said to be incoherent or not coherent.

It has been known for a long time that if you could get incoherent light waves to be in phase, they would contain an enormous amount of energy. But, even though the technology was available earlier, an actual laser was not made until the middle of the 20th century. The principle of operation is not difficult to understand. Recall that in previous discussions in this chapter, it has been noted that whenever an electron gives up energy it is given up in the form of light.

You saw another form of this energy transfer in the x-ray machine discussed in chapter 1. To understand how the first laser—called the ruby laser—operates, it is necessary to know that ruby is a translucent material. Light can pass through it but you cannot read the newspaper through it. If you could it would be transparent. Also, that a *xenon light* is an exceptionally bright light. More important, it can be turned on and turned off almost instantaneously.

Consider now the simple ruby laser illustrated in Fig. 3-13. When the xenon light is flashed on, the energy level of electrons in the ruby is raised. You can think of it as being caused by the electrons taking on additional energy from the light.

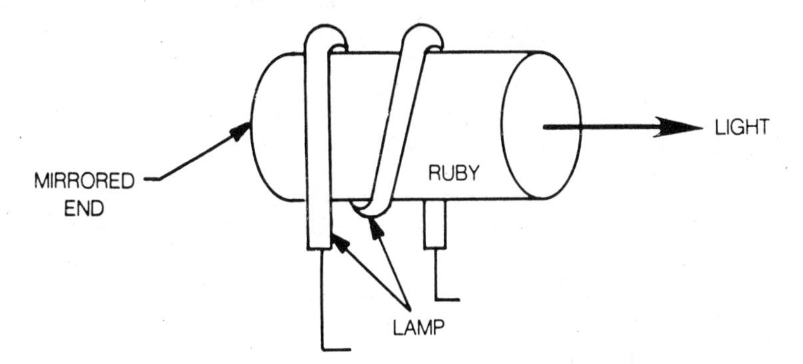

Fig. 3-13. A simplified drawing of the first ruby laser. The lamp is a xenon type.

When the energy level of all of the electrons has been raised, the xenon light is switched off very rapidly. This causes the electrons to go to a lower energy level almost instantaneously. Since they all go to the lower level at the same instant of time, the light waves produced by this change in energy level are in phase, or *coherent*.

One side of the laser tube is mirrored so that the energy is directed out the other side. The amount of energy is enormous when you consider the small size of the laser.

Stories in the newspapers and magazines attest to the high power available with a laser. Those stories include lighting the dark side of the moon with a laser "flashlight," boring holes through metals and other similar applications.

A very important characteristic of coherent light is that it can be focused on a very, very small area. Because of this, the laser can be used in eye surgery and other very delicate operations. Compact disc players that require a very small target area for light of relatively high intensity use diode lasers. Once the laser was made with ruby, other materials were also found to be usable. Gas lasers are very popular.

One very important type of laser is the laser diode. It uses the same basic principle as the LED. The difference is that the electrons cross the forbidden region in a very confined area and the resulting waves are in phase.

Another application of that principle is infrared diodes which are specifically designed to produce infrared energy when the electrons drop to a lower level. The color of light emissions is changed by adding other materials to the gallium arsenide.

ENERGY LEVELS IN A SEMICONDUCTOR JUNCTION DIODE

In the previous chapter you learned about semiconductor diodes and some of their uses in electronic circuitry. We now return to the diode.

Remember that there is a "no-man's land" or "forbidden" region between the cathode and the anode, illustrated in the model of Fig. 3-14.

We now consider the flow of electrons through the diode from the negative side (cathode) to the positive side (anode). The energy level of the electron is shown above the diode in Fig. 3-14. Note that as the electron encounters the forbidden region, energy must be imparted in order to get it through this nonconducting area.

Having passed through the depletion region, the energy level is greatly reduced. The long ramp shown in the energy

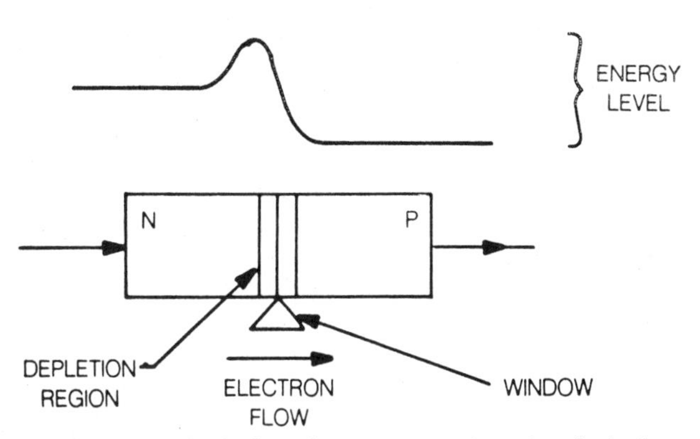

Fig. 3-14. The energy level of an electron moving through a diode drops as it passes through the depletion region.

level is called a potential hill. Following that point, the electrons have a relatively low energy level.

From what was said about the neon lamp, if you suspect that light is given off by the electron when it moves from the higher energy level to the lower level in the diode you would be correct. A "window" diode permits you to observe this light. The device is known as a light-emitting diode (LED). If the diode is made of germanium, there is only about 0.2V difference across the junction. If it is made with silicon, the voltage across the junction is approximately 0.7V. To achieve the brightness levels desirable for light emitting diodes, they are made with gallium arsenide, a combination of the elements gallium and arsenic. This combination produces a material that has many of the same properties as germanium and silicon but the potential difference across the junction of this diode is approximately 1.5V. The greater potential difference produces a greater amount of light when the electrons release their energy.

Light-emitting diodes also behave like light-activated diodes (LADs). In this case, a light is directed at the LED. Photons from the light release electrons from the atoms in the forbidden region. That makes it easier to get an electron through the forbidden region and greatly reduces opposition that the diode offers to forward current flow. LADs are not constructed this way in industry, but the principle is similar. Figure 3-15 shows how an LED can be connected as an LAD.

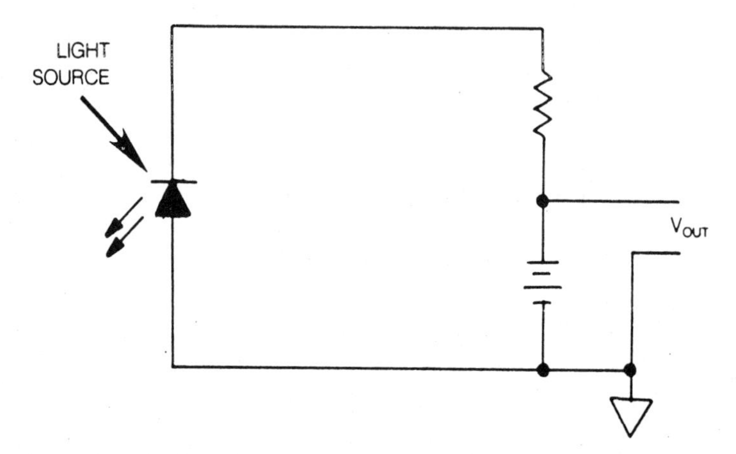

Fig. 3-15. An LED can be used as an LAD. Note that it is reverse biased in this application. This is not a typical use of an LED.

Another strange feature of LEDs is that they behave like zener diodes when reversed biased. This is not a suggested use of LEDs for two reasons. Zener diodes are usually cheaper and the junction of the LED is not designed for continuous reverse voltage applications. However, this property of behaving like a zener diode makes it possible to check LEDs on a curve tracer.

THYRISTORS

Thyristors are electronic devices that have a characteristic breakover voltage starting at zero volts. With a typical thyristor, you can increase the voltage across it until the breakover voltage is reached. Once that happens, the voltage across the device drops to a lower value.

There are electronic tube equivalents to these devices. The neon lamp and the thyratron are examples of tube breakover devices that were used before semiconductor equivalents were made. The concentration in this section is on semiconductor devices, but tube equivalents will be noted as we go through this subject.

Two-terminal Breakover Devices. The most important characteristic of these devices is that they will not conduct electricity until the voltage across them has reached the breakover point. So, they are able to hold off conduction until some desired voltage is reached.

The two most important two-terminal breakover devices are the diac and the Schottky diode.

The Diac. The diac is also called a three-layer diode. Its construction is shown in Fig. 3-16. You might want to think of these as back-to-back diodes.

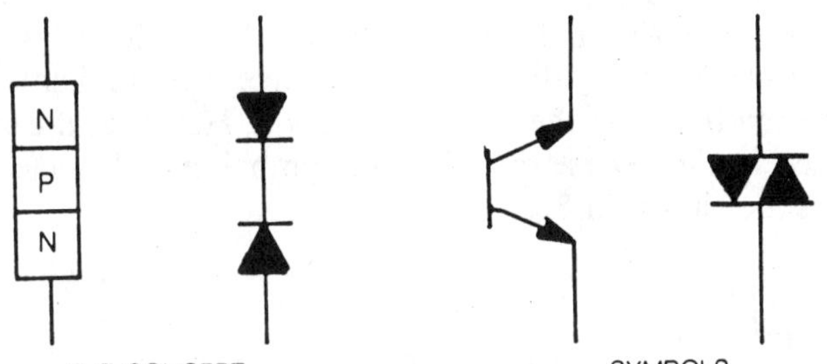

Fig. 3-16. You can think of the diac as being two back-to-back diodes. Front-to-front diodes are shown, which are effectively the same thing.

Diacs are bilateral devices. In other words, once the breakover point is reached, current can flow through them in either direction.

Figure 3-17 shows the characteristic curve of a diac. Note

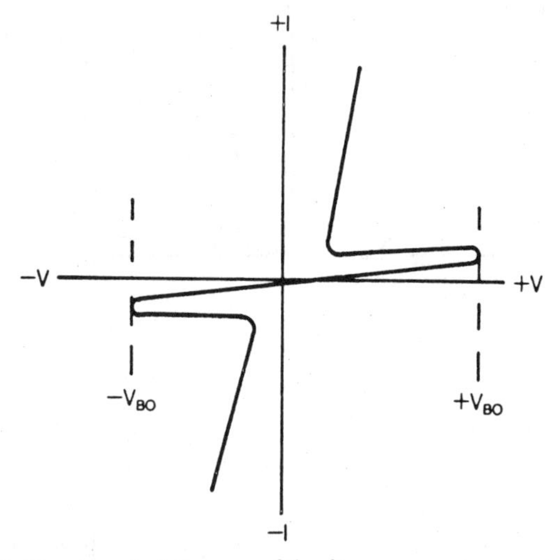

Fig. 3-17. The characteristic curve of the diac.

that once the breakover voltage is reached, the voltage drops to a lower value, and voltage across the device is relatively constant. However, the current through it can increase rapidly to a high value. For that reason, diacs must only be connected into circuits where there is protection against excessive current flow. This can be accomplished by connecting a resistor in series with the devices. Diacs are used extensively in triggering circuits for SCRs and triacs.

The Four-layer Diode. The four-layer diode is also known as a Shockley diode. It has characteristics similar to the diac when biased in the forward direction. However, it cannot be used in the reverse direction. Its symbol and characteristic curves are shown in Fig. 3-18.

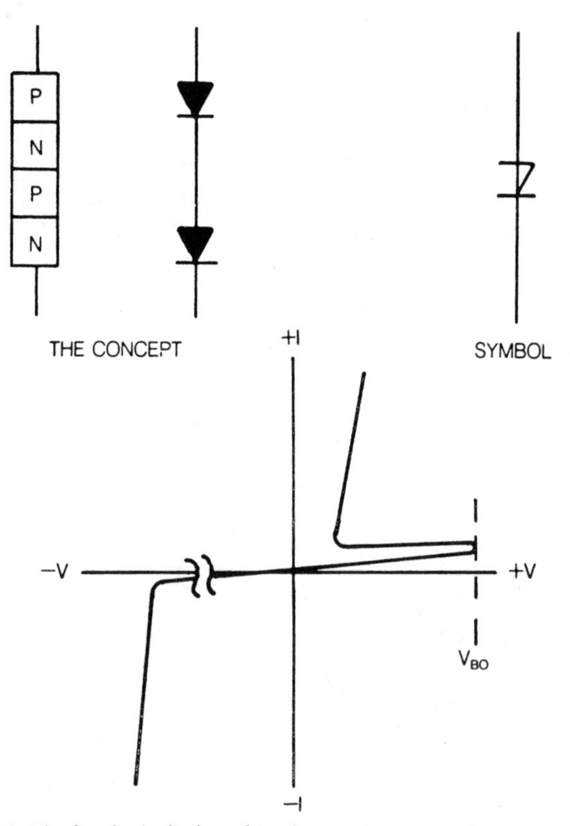

Fig. 3-18. The four-layer diode and its characteristic curve. Compare it with the drawings for the diac.

Three-terminal Thyristors. Three-terminal thyristors are used extensively in industrial control circuits. Their most important characteristic is that their forward breakover voltage can be predetermined by circuit requirements. Also, they can switch very rapidly from a non-conducting to a conducting state.

The SCR. The SCR, or Silicon Control Rectifier, is the solid state equivalent of a thyratron tube. Once you understand the characteristics of an SCR circuit, you also know how a thyratron circuit works. The only difference is in the voltages required for their operation.

Figure 3-19 shows the symbol and a basic thyratron circuit. With a positive voltage on its plate, there would normally be an electron current from its cathode to its plate. However, the current cannot start until a positive pulse is applied to its grid.

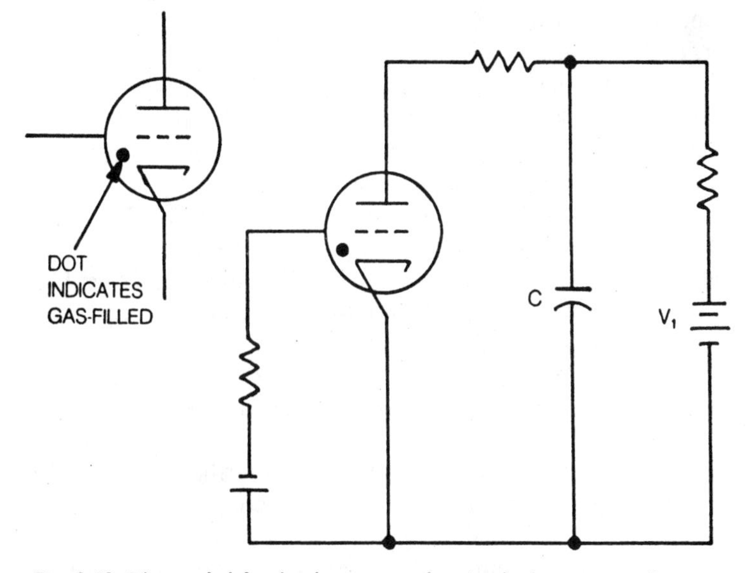

Fig. 3-19. The symbol for the thyratron and a simple thyratron oscillator. V1 sets the breakover point. Capacitor C charges until the voltage across the thyratron reaches the breakover point. This discharges the capacitor. The circuit produces a sawtooth waveform.

Once a cathode-to-plate current is started, it cannot be stopped by the gate. In practice, the plate circuit must be opened or the plate voltage must be dropped to zero volts in order to stop the current. This characteristic is the same as for an SCR.

The symbol and characteristic curves for an SCR are shown in Fig. 3-20. Note that there are several breakover points. For a particular application, the operating breakover point is set by the cathode-to-anode voltage and the gate voltage of the SCR. Once the device conducts, the gate has no further control over it.

The model in Fig. 3-20 is helpful for understanding SCR operation. The PNP transistor provides the base current for the NPN type. Likewise, the NPN transistor provides the base current for the PNP.

Assume that the positive voltage has been applied but the transistors are in nonconducting state. A positive pulse at the gate input starts the NPN transistor into conduction. That forward biases the base of the PNP type and initiates its conduction. Now, each transistor is forward biasing the other transistor, and both are in conducting states.

There is no way to shut this off by applying a negative gate voltage, because the gate voltage supplied by the PNP will override it. This explains why the SCR, once in conduction, cannot be shut off by the gate voltage.

Think about the characteristic curve of the SCR in this way. The gate voltage is increasing from zero to a higher and higher value. At some point the SCR will begin conducting. Up to that point its conduction is minimal and can be disregarded.

Once the SCR starts conducting, the voltage across it is relatively constant but the c· ·rent can become quite high. In fact the current is limited primarily by the components in series with the SCR.

As mentioned before, the gate has no further control over induction once the SCR has started conducting current. Therefore, it is necessary to shut them off using one of several techniques. The most common is to reduce the applied voltage to zero volts. That stops conduction and the gate again controls its operation.

Another method of shutting off an SCR is called commutating. To shut one SCR off, another SCR is used. In some cases, a high-power transistor might be used instead of the SCR.

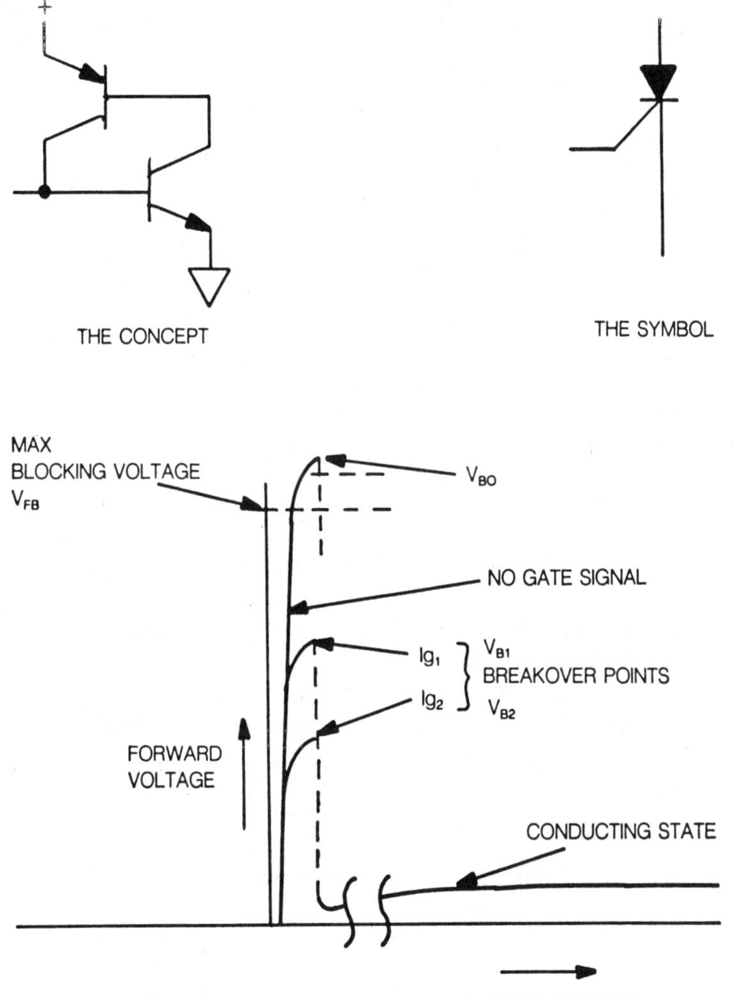

Fig. 3-20. The SCR and its characteristic curve. Note that there are three breakover points. Each is set by the gate current.

The circuit is shown in Fig. 3-21. The overall effect is to produce a negative-going signal on the anode. That is sufficient to drive the SCR into cutoff.

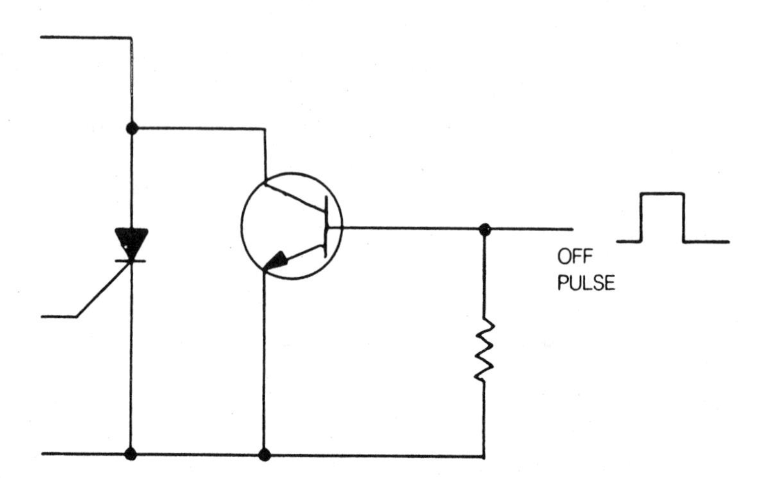

Fig. 3-21. The transistor in this connection is used to shut off the SCR.

OPERATING CHARACTERISTICS OF SCRs

The basic characteristics of SCR and triac operation have been discussed. However, the best way to understand these devices is to understand how they work in some simple circuits. We will now review these circuits.

The SCR in a Simple DC Circuit. Figure 3-22 shows an SCR connected into a simple dc circuit. The applied voltage (V)

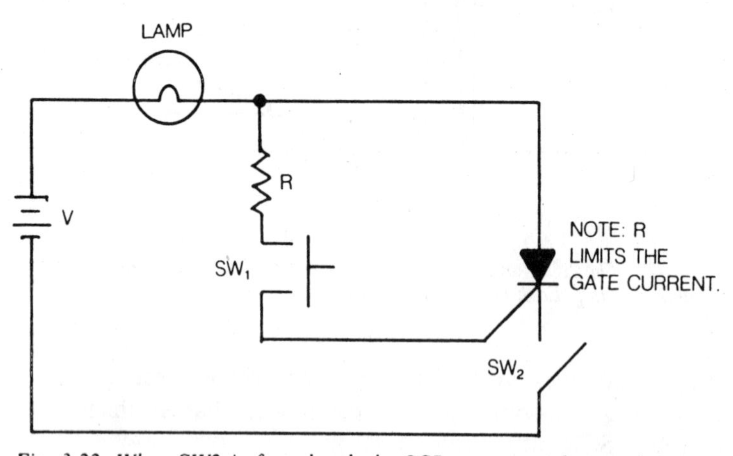

Fig. 3-22. When SW2 is first closed, the SCR cannot conduct until SW1 is momentarily closed. The lamp cannot be turned off except by SW2.

must be sufficient to light the lamp and produce the forward voltage drop across the SCR when it is in conduction. That forward drop is often negligible in comparison with the lamp voltage.

We will start with both switches open and the lamp in the off condition. SW2 is closed first, which completes the circuit, but the lamp does not light because the SCR has not been triggered into the on condition.

Momentarily closing SW1 provides the necessary gate voltage to cause the SCR to conduct. The lamp lights then even though the voltage on the gate has been removed by releasing SW1. Repeated triggering of the SCR by SW1 has no effect on the on condition of the lamp. The only way to get the lamp off is to open switch SW2 in this circuit. Circuits have been devised to turn the lamp off. A basic application will be shown in this section.

Figure 3-23 shows another SCR circuit. In this case, the applied voltage is the secondary of a transformer (L_s). Since the voltage is ac, it drops to zero volts every half-cycle. With SW1 open, the SCR does not conduct.

Assume that SW2 is closed. Then closing switch SW1 momentarily will provide a positive pulse on the next positive half-cycle of input voltage. That turns the SCR on.

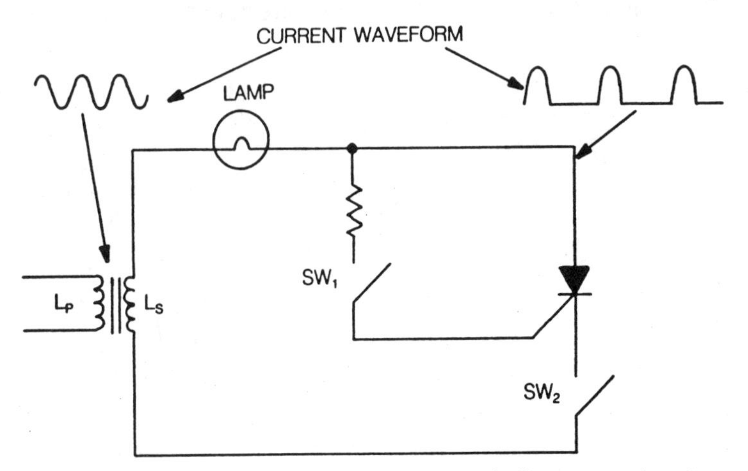

Fig. 3-23. In this ac application, the SCR is turned off every time the voltage across L_s drops to zero.

As long as switch SW1 is closed, a positive gate voltage will be delivered on each positive half-cycle so the SCR will conduct.

On the half-cycles when the voltage on the anode of the SCR is negative, no conduction takes place. Remember that this is, after all, a diode. It can only conduct when its anode is positive with respect to its cathode. The waveform for the current through the lamp shows that the applied voltage produces current only on its positive half-cycle.

Figure 3-24 shows how to use an SCR in a full-wave circuit. The bridge rectifier across the transformer secondary converts the output pulses to positive half-wave cycles. Since this is a full-wave rectifier, there is an output for both half-cycles. Therefore, when the switch is closed, the lamp will light at full brightness. The lamp in Fig. 3-23 can only light at half brightness because there are half-wave pulses flowing through it. This technique of using a bridge rectifier to produce full-wave operation is often used in industrial electronics circuits.

You may wonder if there is any practical application to any of these circuits. Consider the problem of turning the lamp on at some remote position. The circuit to be used must have a low current requirement and it must produce positive action. In reality this is one of the functions of a relay. So, you could consider the ac circuit of Fig. 3-23, and the dc circuit of Fig. 3-24 to be electronic relays. A very low gate power is required to

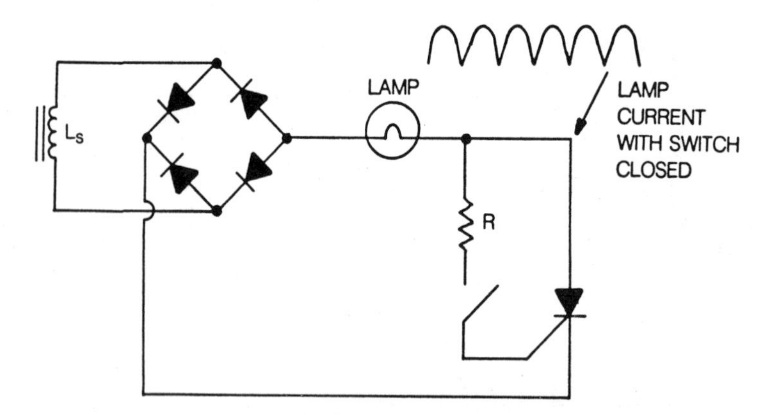

Fig. 3-24. By using a bridge rectifier you can get full-wave operation with an SCR.

turn on a high-powered SCR, so the circuit is useful for remote switching.

The Snubber. An inherent problem of the SCR is that it can be turned on by cathode-to-anode noise voltages that may come on at a time that is very inconvenient or undesirable. A circuit called the snubber is placed across the SCR to prevent this. It is shown in Fig. 3-25. If a transient pulse occurs across the SCR as a result of noise introduced into the system along the power line, the snubber acts as a filter to pass the pulse around the SCR. This prevents an undesirable on condition.

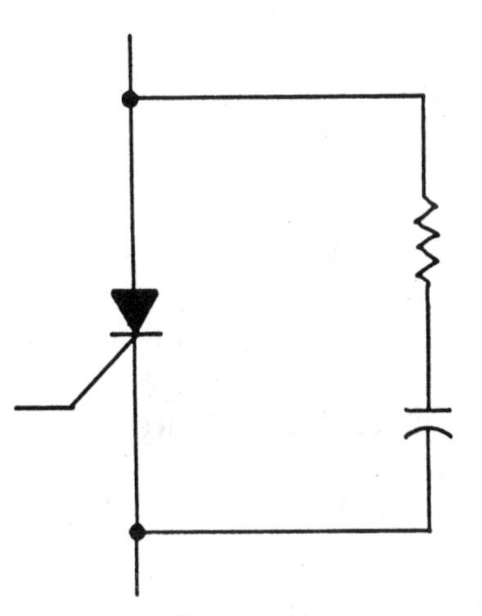

Fig. 3-25. The RC circuit is called a snubber.

The Triac. The triac is very much like the SCR. However, the SCR is a unilateral device. You can tell this by the diode symbol that is included in its overall symbol. It will conduct electrons from cathode-to-anode but not in the reverse direction.

In the case of the triac, conduction can take place in either direction. Study the symbol and observe that it looks like back-to-back diodes. In reality, it is back-to-back SCRs. This is shown in Fig. 3-26. In the early days before triacs were popular, engineers would sometimes connect back-to-back parallel SCRs to get bilateral conduction.

The Silicon Controlled Switch and the Gate Turn On Device. You will remember that an SCR can be switched on

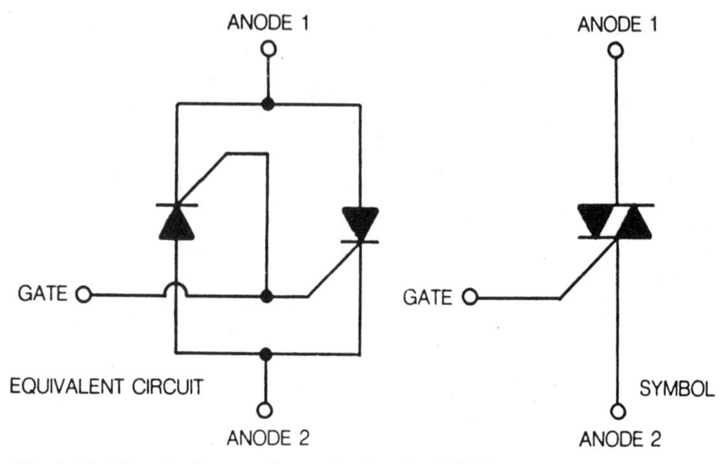

Fig. 3-26. The triac is actually two back-to-back SCRs.

but cannot be switched off. That is the difference between an SCR and an SCS. With an SCS, the device can be switched on with an input voltage, and it can also be switched off. It is shown schematically in Fig. 3-27. If you compare the model with the SCR you will see that the devices are very similar. The main difference is the addition of a cathode gate and anode gate as opposed to a single control gate in the SCR.

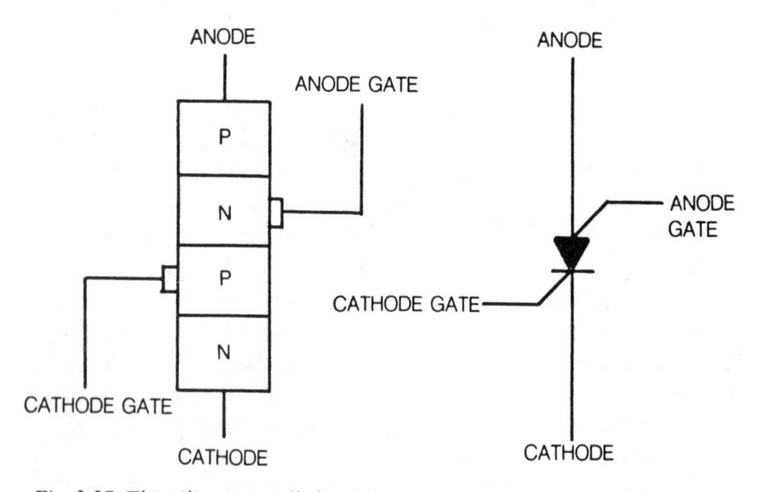

Fig. 3-27. This silicon controlled switch can be used to start or shut off an anode current.

As you might expect, it takes more current to switch the device off at the anode gate than it does to switch it on at the cathode gate. The actual current required is dependent upon the device itself and also upon such things as ambient temperature and voltage across the device.

Since the silicon control switch can be turned off much more rapidly than an SCR, that is one of its advantages. However, you do not find silicon control switches in the high-power ratings obtainable with an SCR.

Gate Turn-Off. The model and schematic symbol for a gate turn-off device is shown in Fig. 3-28. Again you will notice the similarity to an SCR and an SCS. The difference between this device and the other two devices is that it can be turned on or off with a voltage at one gate.

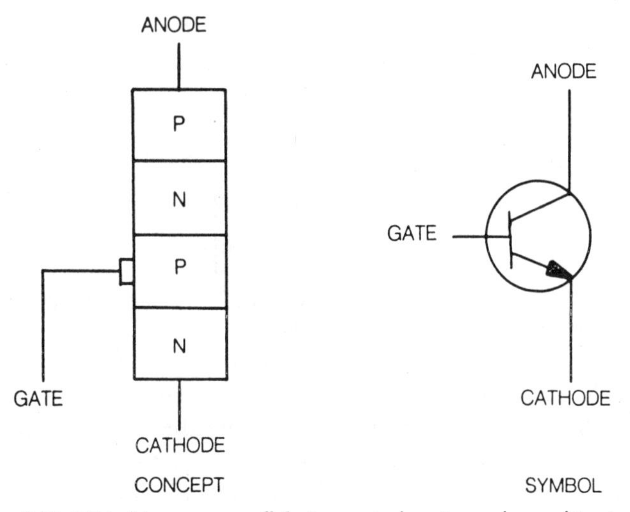

Fig. 3-28. With this gate turn-off device, a single gate can be used to start and stop the anode current.

The Unijunction Transistor. Figure 3-29 shows a symbol for a device called a unijunction transistor. It gets its name from the fact that it has one emitter and two emitter-to-base junctions, but no base-to-collector junction.

The operating characteristics for this device are shown in Fig. 3-29. it does not conduct current until the emitter voltage is some proportion of the base or applied power-supply voltage.

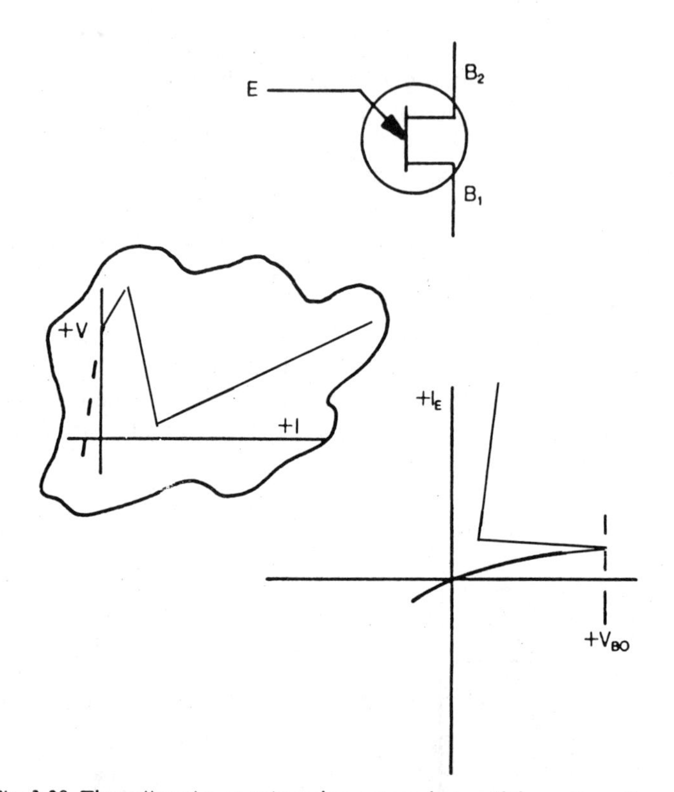

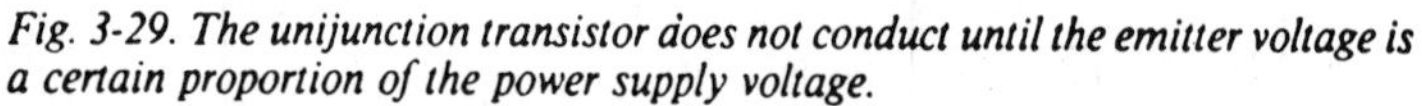
Fig. 3-29. The unijunction transistor does not conduct until the emitter voltage is a certain proportion of the power supply voltage.

This component is best understood when it is connected into a simple oscillator circuit like the one shown in Fig. 3-30. Assume that the switch has just been closed and the capacitor begins to charge through the resistor. The amount of time it takes to charge to the required emitter voltage is dependent upon the time constant (R × C).

When the proper emitter voltage is reached, there is conduction in the device. Of specific interest here is the fact that there is an emitter-base conduction across the capacitor. This conduction discharges the capacitor. Once the capacitor is discharged, the emitter voltage is below the value required for conduction and the capacitor begins to charge again.

There are two resistors, one in the base one and one in the base two circuits. During the period of time when the UJT is

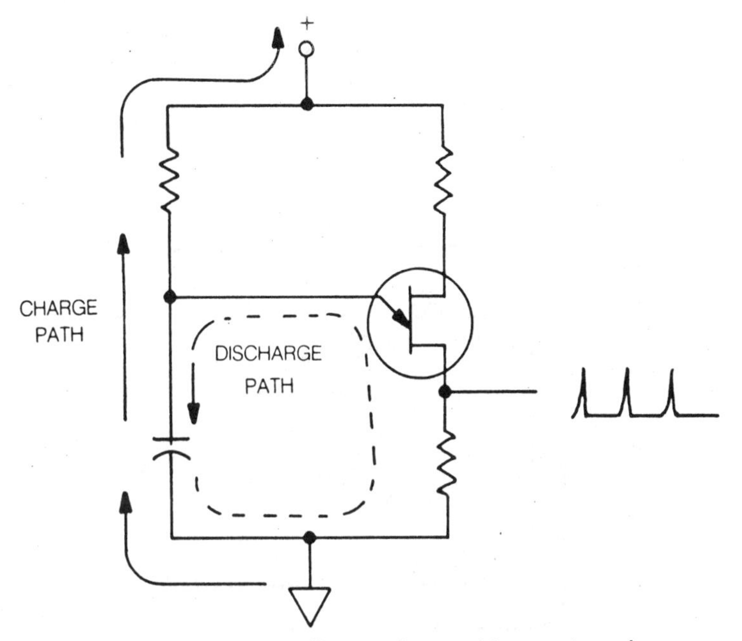

Fig. 3-30. This simple UJT oscillator produces positive outgoing pulses.

conducting to discharge the capacitor, there is also current through the device from phase one to phase two. This produces a voltage across the base resistors. That voltage is a short-duration pulse because it does not take long to discharge a capacitor.

The capacitor does not discharge all the way to zero volts. It discharges to a voltage below that required for conduction of the UJT. Normally, it is the output pulse at base one that is used.

The emitter of the UJT must be some proportion of the applied voltage in order for the UJT to conduct. That proportion is called the intrinsic standoff ratio.

As an example, an intrinsic standoff ratio of 0.63 means that the emitter voltage must be 63 percent of the applied voltage before the UJT can conduct. The intrinsic standoff ratio is set during the manufacture of the UJT.

The Programmable UJT (PUT). Figure 3-31 shows an oscillator circuit using a programmable unijunction transistor. In this case, the time constant circuit is comprised of R1 and C. Resistors R_a and R_b form a voltage divider to set the control

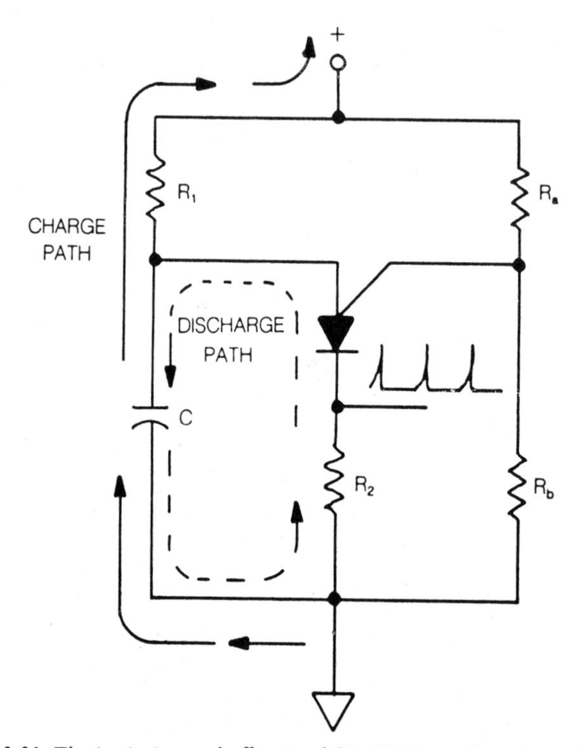

Fig. 3-31. The intrinsic stand-off ratio of this PUT is set by voltage divider R_a and R_b.

electrode voltage. That, in turn, sets the intrinsic standoff ratio of the PUT. Once that ratio is set, the oscillation time is fixed. Now capacitor C begins to charge through R1 until the voltage across the device is sufficient to cause it to conduct. When it conducts, there is an output pulse across resistor R2. At the same time, capacitor C discharges.

As with the unijunction transistor, that capacitor does not completely discharge. For this reason, the time for discharge is considerably shorter than one time constant. The output pulse is a short duration voltage across R2. The charge and discharge paths for the capacitor are drawn on the circuit of Fig. 3-31.

Optoelectronic Thyristors. Instead of using a gate voltage to turn the SCR on, an SCR can be triggered by a light source. In that case they are called light-activated SCRs or LASCs. The presence or absence of light is used for gate turn-on but these devices also have a gate lead so it can be turned on, or a turn-on threshold can be set by the gate voltage. As with other semiconductor switching devices, the applied voltage and also the amount of ambient temperature are factors in the amount of light required for turning the LASC on.

MULTIPLEXERS AND DEMULTIPLEXERS

A multiplexer is a component or device that permits many inputs to be switched, one at a time, to a single output. The principle of the multiplexer is shown schematically in Fig. 3-32. In their original form, multiplexers were electromechanically operated. They were called stepping switches. Selection was made by applying a pulse for each change in the switching contacts. Today, multiplexers are made in the form of integrated circuits, but they perform the same job.

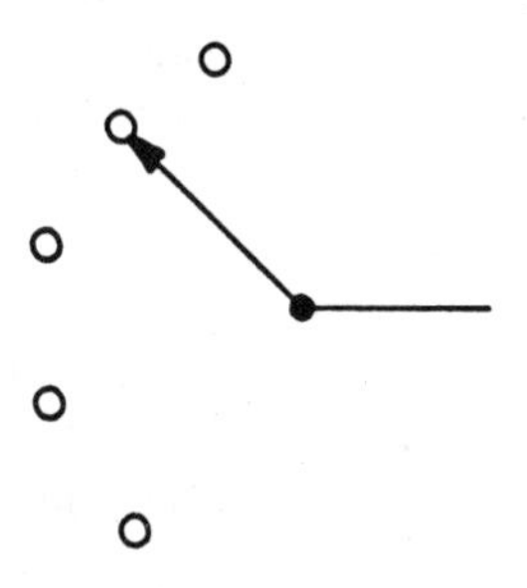

Fig. 3-32. Multiplexers are used to select one of a number of inputs.

Demultiplexers perform the opposite job. They take a single input and switch it to various outputs. A demultiplexer is shown schematically in Fig. 3-33. This type of device was originally made with a stepping switch. It is accomplished today by demultiplexers packaged in integrated circuit form.

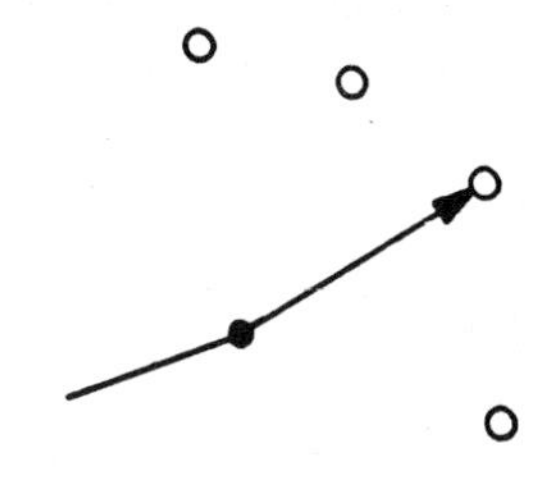

Fig. 3-33. A demultiplexer delivers a signal to one of a number of outputs.

SUMMARY

The subject of this chapter is switching. It includes discussions of devices that are used for switching and devices that are operated by switching. Switches are not necessarily on/off devices. For example, a switch may be used to select between two voltage levels.

Relays are remotely operated electromechanical switches. Today there are electronic relays which have no moving parts, which are favored in many applications. For example, an electronic relay can be used in an explosive atmosphere with no danger of arcing at the switching contacts. An electromechanical relay normally would not be good for that application, except that reed switches, with their contacts sealed can match that advantage of electronic relays.

Another disadvantage of the electromechanical relay is moving parts, which make it unsuitable for use in applications where there is a lot of vibration and pounding.

Another disadvantage of relays is contact bounce when the relay is energized. This makes relays unsuitable for use in integrated logic circuits of the type you will be reading about in chapter 4.

Lasers are included in this chapter because they produce light by switching the energy levels of electrons. They are easily understood once you have reviewed the basic principles of neon lamp operation.

Thyristors are breakover devices. They will not conduct until a predetermined condition exists. In the case of two-terminal thyristors, the voltage level must be a certain minimum value before they conduct.

In three-terminal thyristors, conduction is dependent upon the voltage or current of a control electrode. In the case of the SCR it is gate current that operates the device.

A silicon-controlled switch is a four-terminal thyristor. It can be turned on and off with voltages.

There are two devices that have their conduction set by an intrinsic standoff ratio—that is the ratio of the control electrode voltage to the power supply voltage.

In the UJT, the intrinsic standoff ratio is set during manufacture. With a PUT, a voltage divider sets that ratio. You can determine the intrinsic standoff ratio for a PUT by the ratio of resistances in a voltage divider. For that reason, the device is said to be programmable.

Multiplexers and demultiplexers are used extensively in modern industrial circuits. In the early days their job was accomplished by stepping switches. Today, there are semiconductor equivalents for those switches.

SELF TEST

1. A laser light has:
   (A) coherent waves.
   (B) inherent waves.

2. In what component would you expect to find Crookes dark space?
   (A) lasers.
   (B) neon lamps.

3. A proximity circuit might use:
   (A) a VDR switch.
   (B) a reed switch.

4. You have to use your expertise as a technician to know when a symbol for an open contact is not:
   (A) a capacitor.
   (B) an SCR.

5. In a certain relay circuit, a coil current closes the normally-open contacts. This is a:
   (A) Form A configuration.
   (B) Form B configuration.

6. Which of the following is similar in operation to a diac?
   (A) neon lamp.
   (B) thyratron.

7. An SCR circuit can behave like:
   (A) a relay.
   (B) an amplifier.

8. Which of the following could be an intrinsic standoff ratio?

(A) 6.3.
(B) 0.63.

9. The output waveform of a UJT oscillator at base one is a:
   (A) sawtooth.
   (B) short-duration pulse.

10. Could a neon lamp be used as a voltage regulator?
    (A) Yes.
    (B) No.

ANSWERS TO SELF TEST

1. (A) Coherent light waves are in phase. They carry very high power.

2. (B) When electrons leave the surface of the metal they must travel a short distance before combining with a positive ion. Across that distance there is no light.

3. (B) The switch operates when a magnet is brought close.

4. (A) The capacitor symbol should have a curved line to represent one of the plates. However, you may see two parallel lines being used, and the only way to know if it is a capacitor or switch is to rely upon your knowledge of electronics.

5. (A) The contact forms are shown in Figure 3-6.

6. (A) Both the neon lamp and the diac can conduct in two directions. In other words, they are bilateral. Neither will conduct until the voltage across it reaches a certain minimum value.

7. (A) An SCR circuit that behaves like a relay was shown in this chapter.

8. (B) The ratio is always less than 1.0. A value of 0.63 is sometimes used, so that the oscillation frequency is approximately equal to the reciprocal of the time constant.

9. (B) The pulse occurs while the capacitor is discharging. This is a very short time compared to the charging time for the capacitor. The short period of discharge current flowing through R produces a short-duration pulse.

10. (A) The characteristic curve of the neon lamp shows that the voltage across it is constant for a specific range of currents. A disadvantage of the neon lamp is that it generates electrical noise.

4

Digital Logic Gates and Microprocessors

SINCE IT IS impossible in one chapter to cover all aspects of digital logic and microprocessors, we will concentrate on what is most commonly encountered by industrial electronics technicians. In the last chapter, some troubleshooting techniques for digital logic systems will be included.

Industrial electronic systems are expensive and it can be very costly to a company if they are out of operation for any even a short period of time. For that reason, operation and troubleshooting manuals are available for those types of equipment.

You should make it a point to read those manuals and study them as part of your preparation for your job. The manuals are written so you can understand the basics of digital logic included in this chapter.

CHAPTER OBJECTIVES

After reading this chapter you will be able to answer the following questions:

- What are the basic gates and their truth tables?
- How do you read timing diagrams?
- What does a toggle flip flop do?
- What is the advantage and the disadvantage of a synchronous counter over a ripple counter?
- What does a microprocessor do and how is it used in industrial electronics?

THE BASIC GATES

When you first studied analog electronics, you had to learn components such as inductors, capacitors, and resistors. These components are the building blocks of circuits and systems. You did the same thing when you are first starting to study digital logic. Instead of the components used in linear circuits, the basic building blocks in logic systems are gates. You may have studied the gates before, but it is a good idea to review them occasionally.

In this discussion I have included some ladder diagrams to give you practice in reading those types of schematics. Also, you should have the truth tables memorized. They are reviewed in this chapter.

If you are quickly and efficiently troubleshooting a logic system, you cannot stop to look up a truth table of the logic gate to determine if it is working properly. As with analog systems, if you are efficiently troubleshooting an enhancement P-channel MOSFET, you cannot stop to look up the dc operating voltages to determine if the component is properly biased. You need to know that by memory.

Inverter Gate (NOT Gate). As its name implies, the output of the inverter is opposite to its input. In digital logic systems we only have two levels of signal: logic 0 and logic 1. In most operations logic 0 will be near zero volts and logic 1 is between +3.5 and +5 volts. So, if the input is logic 1 to an inverter, the output will be logic 0. Conversely, if logic 0 is the input, the output will be logic 1.

The concept of the inverter is reviewed in Fig. 4-1. The NPN transistor circuit shows an inverter gate can be made with a conventional amplifier. The input signal is to the base and the output is from the collector, so this is a common-emitter amplifier. This means that the output is 180 degrees out of phase with the input.

With the switch closed (or at logic 1), the transistor is properly biased and the transistor conducts heavily. The voltage drop across the collector load resistance is so large that the lamp does not light. This means the lamp is in a logic 0 condition when the switch is in a logic 1 condition.

When the switch is in the logic 0 position, there is an open in the base bias circuit. This means that the transistor is cut off. There is no voltage drop across the collector load resistance due to transistor collector current. Therefore the lamp will be lighted.

The conventional relay diagram in Fig. 4-1 shows how an inverter can be made with that component. When the switch is open, the relay is de-energized. Remember that relays are always shown in their de-energized position. In this illustration the lamp will be on because it is directly across the battery (V).

When the switch is closed, the relay is energized and the armature goes to the unconnected electrode. This opens the lamp circuit and the lamp goes off. The switch positions are labeled 0 and 1. When it is in the 0 position, the lamp is in a

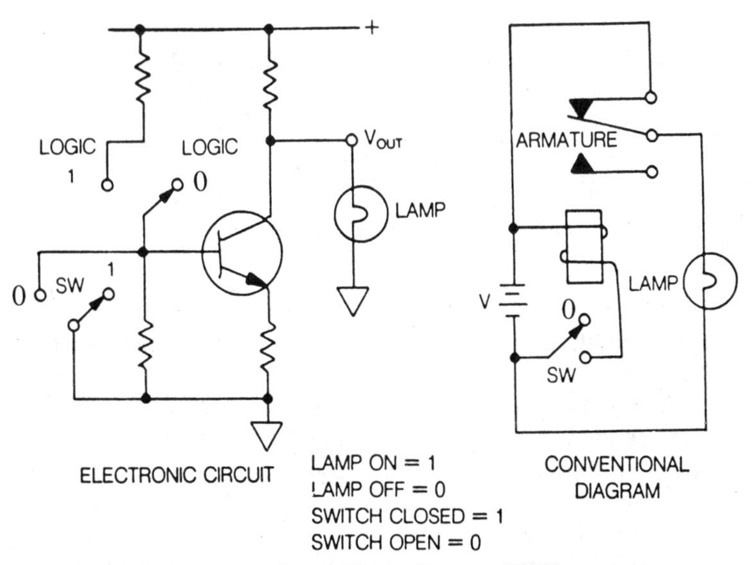

Fig. 4-1. Some important characteristics of inverter (NOT) gates.

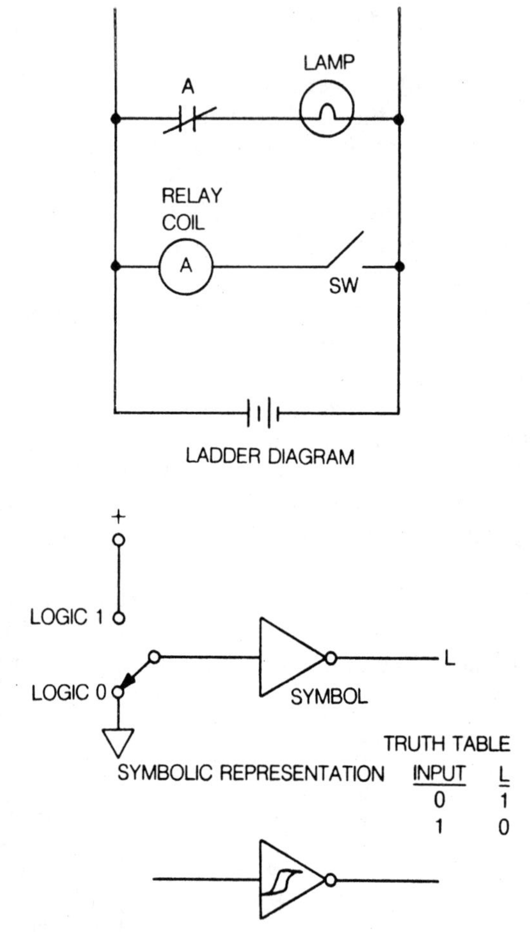

Fig. 4-1. cont

logic 1 condition. When a switch is turned to logic 1, the lamp is off (logic 0).

Many relay circuits are shown in industrial electronic diagrams as ladder diagrams. An example is shown in Fig. 4-1. This type schematic is easier to draw and requires less space than conventional drawings. This is especially true in complicated circuits.

Ladder diagrams are also used in other applications in

industrial electronics. For example, a form of ladder diagram is used for programming programmable controllers. A programmable controller is a very versatile microprocessor system used for controlling machines or industrial processes. It can be set up so the machine or process can be used in various applications, simply by programming the electronic (or relay) system.

Originally, programmable controllers were made with relays. In order to prevent confusion with the new electronic controllers, they were set up so the person knowing ladder diagrams could program them.

In the ladder diagram of Fig. 4-1, the coil is shown in series with a switch. A standard symbol is used for the coil in ladder diagrams. The letter "A" means that any place in the ladder diagram that has an "A" by the contacts will be operated when this coil is energized.

The relay contact in series with the lamp represents a normally closed switch contact. When the relay is not energized, the normally closed contact permits current to flow through the lamp and the lamp is on. When the relay is energized, the normally closed contact opens and the lamp goes off. The ladder diagram is for the same circuit shown in the conventional relay diagram. Note that the battery is connected directly across the two ends of the ladder. In many applications the power source is not shown.

You are more likely to find the NOT or Inverter circuit represented as shown in the symbolic representation of Fig. 4-1. In this illustration, a switch is used to show that the input can be either logic 1 or logic 0. In most applications the input will come from some other logic gate. Observe the symbol carefully. The circle always represents inversion in logic systems that use these symbols.

The truth table for the logic inverter is also shown in Fig. 4-1. It shows that the output logic level (L) is always opposite to the input.

Figure 4-1 also shows an inverter that uses a Schmitt trigger circuit. The symbol inside the inverter is called a hysteresis curve. This symbol on a component means that it employs a Schmitt trigger circuit.

Basically, a Schmitt trigger is a feedback circuit that employs speed-up capacitors to provide very fast switching from a low condition to a high condition, and from a high to a low. The Schmitt trigger inverter can be used as a replacement for other inverter applications, provided the very fast triggering does not introduce timing problems in the overall system.

Tri-State Inverters and Buffers. Figure 4-2 shows a tri-state buffer and a tri-state inverter. These devices are included here because of their similarity to the inverter symbol.

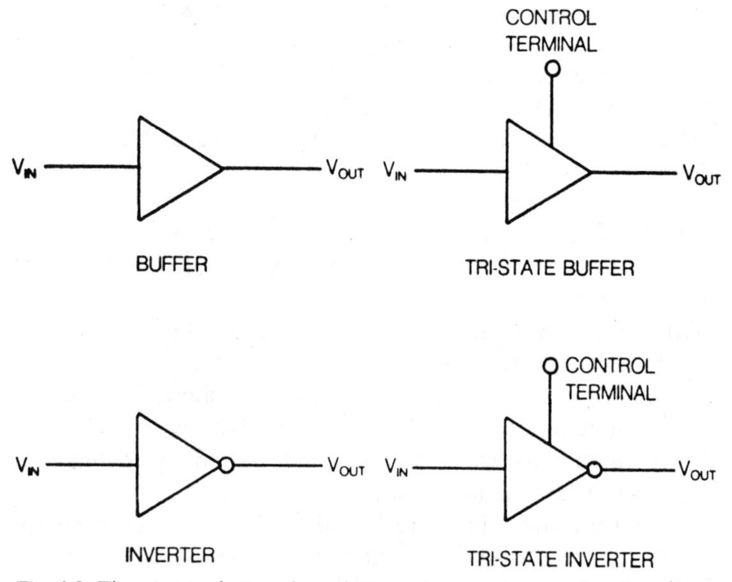

Fig. 4-2. The tri-state devices shown here can be turned on or off with a voltage on the control terminal.

A *buffer* is simply an interface between two circuits. It may provide current gain, but seldom provides voltage gain because the maximum voltage is at logic 1, which is usually 5 volts. A buffer can prevent one circuit from affecting another. For example, the output of an oscillator may go through a buffer to prevent the circuitry from pulling the oscillator off frequency.

The tri-state buffer has a third input that permits the buffer to be turned on and off by a voltage signal. An inverter can also be used as a tri-state device. As the symbol in Fig. 4-2 shows, an additional electrode is added to the inverter so it can be turned on and off by a voltage signal.

The devices shown in Fig. 4-2 are often built into complicated logic systems such as microprocessors. They can be used to prevent data on the line from feeding back into a logic system, which could be destructive in some cases. Suppose, for example, there is a logic 0 at the input of the inverter. That means that the output is at logic 1, or plus 5 volts. If, during the operation of the system, the line on the output of the inverter is driven to logic 0 by some other system, there will be a simultaneous logic 1 (5 volts) and a logic 0 (0 volts). This is a short circuit. It would likely result in destructive currents.

If the tri-state inverter is used and the input electrode is in the logic 0 condition, the output line will be disconnected. Therefore, voltages on the output line cannot produce a short-circuit condition.

The OR Gate. Figure 4-3 shows the characteristics of the Or gate. The simple schematic diagram shows that the light will be on if either A or B is closed—that is, either or both are in a logic 1 condition. Both switches must be open in order to get the lamp off, or in a logic 0 condition.

The relay diagram shows that the coil can be energized by either switch A or B. The lamp will be on when either or both of those switches are closed.

The ladder diagram is the same schematically as the conventional diagram. Note that the relay contact is now a symbol for a normally open contact. When either A or B (or both) are closed, contact A will close and the lamp will be on.

You will most often see the Or gate represented symbolically. For the illustration in Fig. 4-3, A and B are controlled by switches that permit either a 1 or 0 to be delivered to each of the input leads. The truth table shows the results. If both switches are open, the lamp is off. If either or both is closed the lamp is on.

Logic AND Gates. Figure 4-4 shows the characteristics of a logic AND gate. The simplified diagram shows that both A and B switches must be closed before the lamp (L) can be on, or in a logic 1 condition. The relay circuit shows that the coil can be energized by either of two switches.

The ladder diagram, which is the equivalent of a conventional diagram, again shows a normally open contact. It is

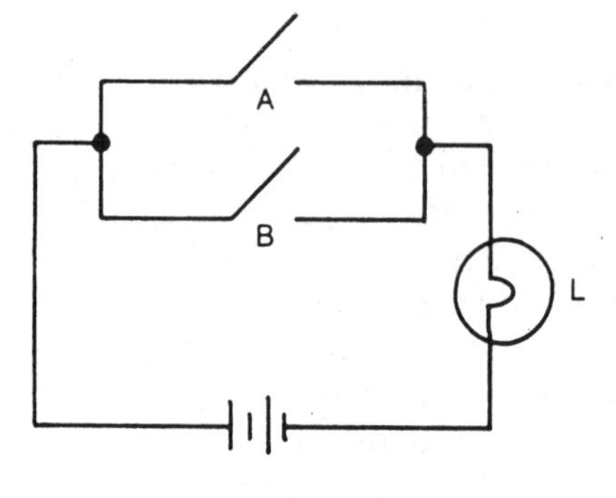

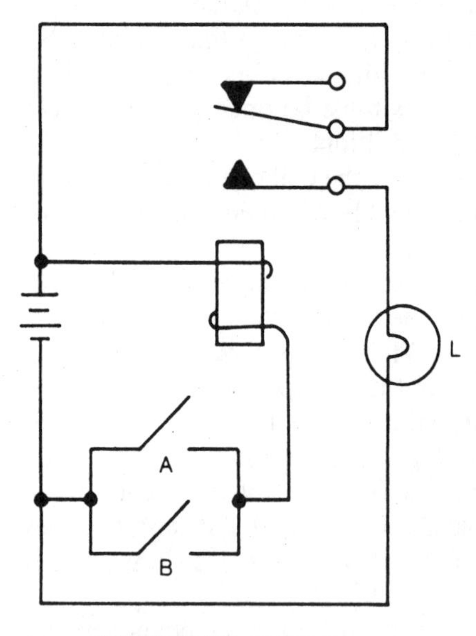

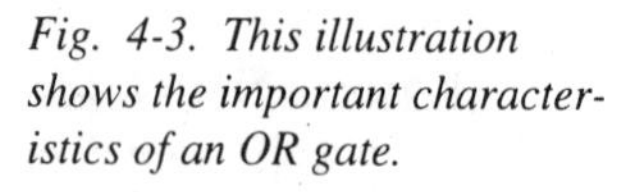

Fig. 4-3. This illustration shows the important characteristics of an OR gate.

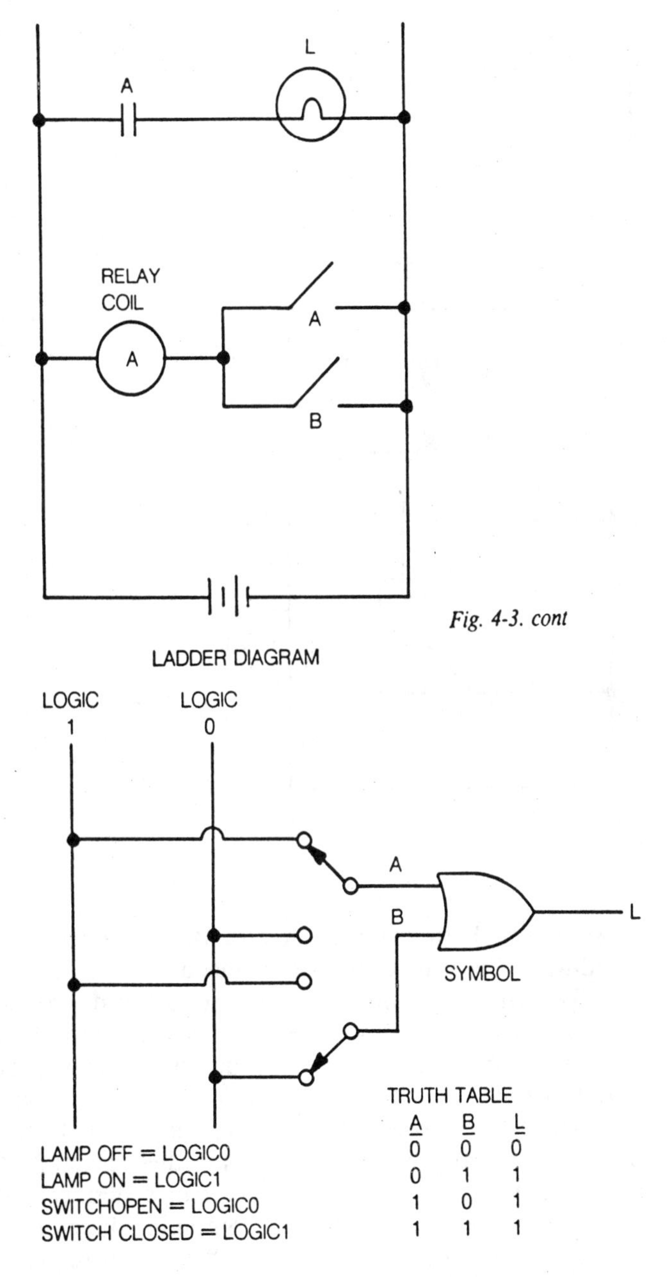

A	B	L
0	0	0
0	1	1
1	0	1
1	1	1

Fig. 4-3. cont

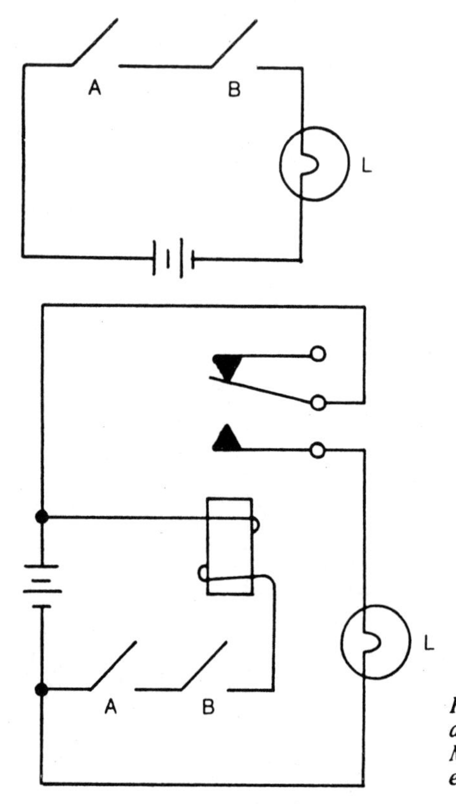

Fig. 4-4. These AND gate characteristics are very important. Make sure you understand each one.

closed when both switches (A and B) are in the logic one condition. This state closes the contact and turns the lamp on.

The symbolic representation is set up to deliver either a logic 1 or logic 0 to the two contacts. The truth table shows that both inputs must be in a logic 1 condition in order to get a logic 1 out. All other input switch conditions produce a logic 0.

These three basic gates are very important. Every other logic system can be constructed using these gates. In fact, a complicated computer or microprocessor system can be constructed, at least theoretically, by combining these three gates.

In the real world you will not often see these gates made from individual components. In today's systems they are usu-

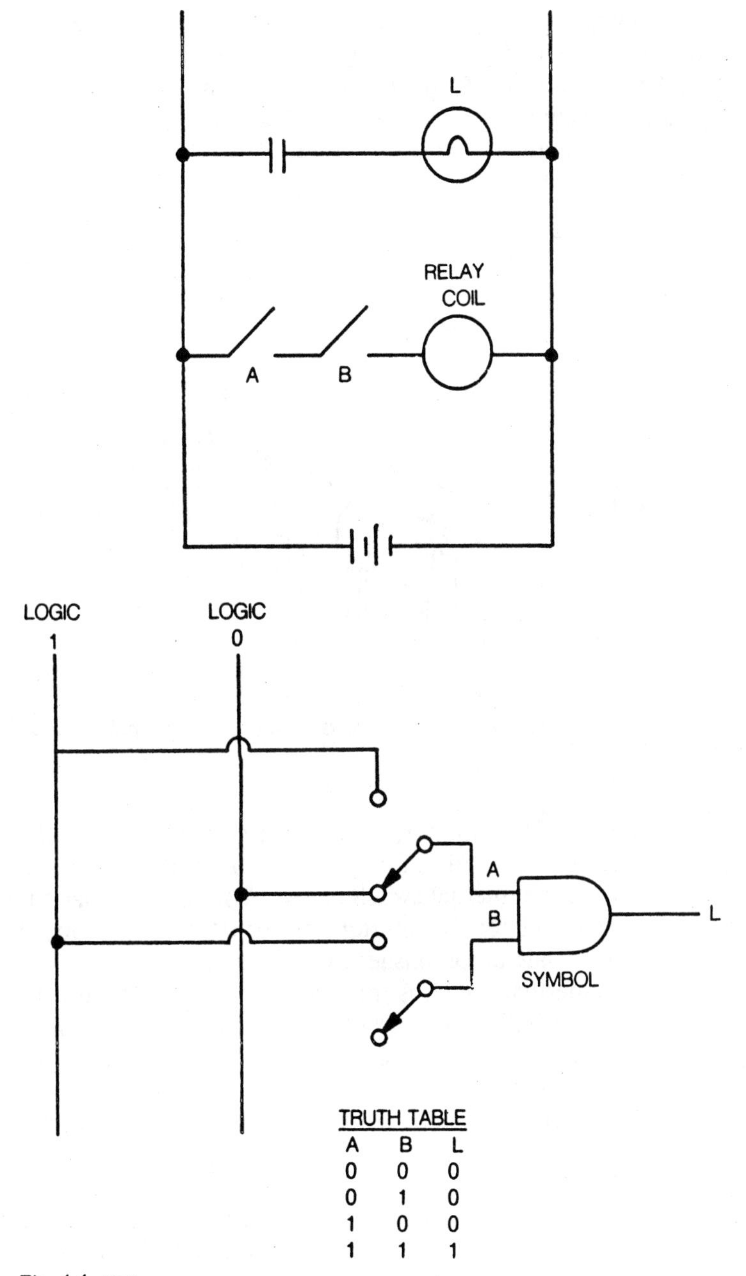

TRUTH TABLE

A	B	L
0	0	0
0	1	0
1	0	0
1	1	1

Fig. 4-4. cont

ally in integrated packages like the one in Fig. 4-5, or they are part of a more complicated integrated circuit.

Using the Logic Gates Discussed So Far. If you have a good understanding of the inverter, OR, and AND gates, you will be able to answer these questions. (Answers are at the end of this section.)

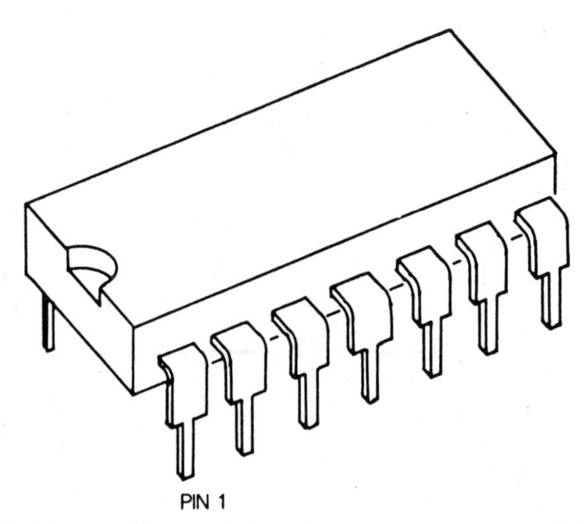

Fig. 4-5. This type of integrated circuit houses many gates. It is called a dual in-line PIN (DIP) package.

1. In a CRT monitor, an internal switch automatically turns the circuit off. So, to get it on, it is necessary to have the internal switch closed. Also, the external on/off switch for the monitor must be closed. What type of logic gate accomplishes this function?
2. A thermal switch is mounted on a motor to turn the motor off if it overheats. Otherwise, the motor is turned on and off by a manual switch. Which logic gate accomplishes this?
3. Two input logic 0 signals are required to turn a machine off in the middle of a run. What logic gate accomplishes this?
4. A lamp on a panel comes on if, for some reason, a certain motor is stopped. Otherwise, the lamp is off. What logic gate accomplishes this?

5. It is necessary to turn a lamp on from either of two positions at the same time. It is also necessary to be able to turn the lamp off from the same two positions. Is this function represented by an OR of the type shown in Fig. 4-3?

Answers:

1. AND
2. AND
3. OR
4. NOT or Inverter
5. No (It is an Exclusive OR—to be discussed later in this chapter.)

NORS and NANDS. Figure 4-6 shows an OR gate and an AND gate with an inverter at the output. These are called NOR and NAND gates. Compare the output (L) of each gate with the OR and the AND. Their output is equal to the output of those gates inverted. So, in the NOR gate, two input zeros produce an output of logic 1. For the OR gate, two input zeros represent an output of 0. For the NAND, two input zeros represent an output of logic 1. But, in the AND, two zeros represent an output of logic 0.

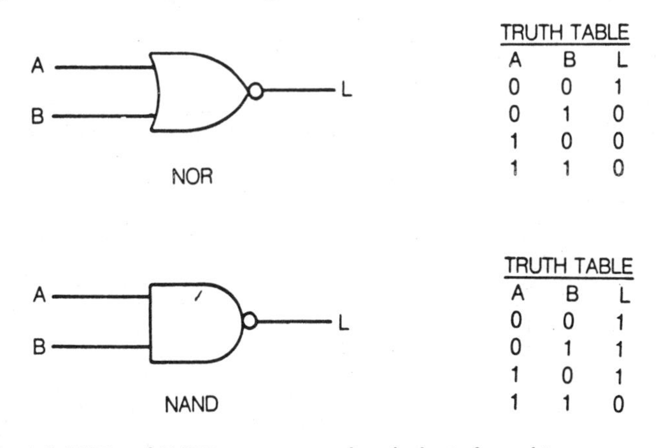

Fig. 4-6. NOR and NAND gates are used as the basis for making more complicated logic circuitry. Be sure you understand the truth table for these important gates.

The two gates in Fig. 4-6 are very important. They are used as basic construction units in many integrated circuit families. Families of integrated circuits have the same power supply requirements, similar propagation delays (that is, delays of signals passing through the gates) and other similarities. Important families are TTL, CMOS, and ECL (Emitter Coupled Logic).

Exclusive OR and Exclusive NOR Gates. Figure 4-7 shows the symbols for Exclusive OR and Exclusive NOR gates. The Exclusive NOR gate is also referred to as a Logic Comparator.

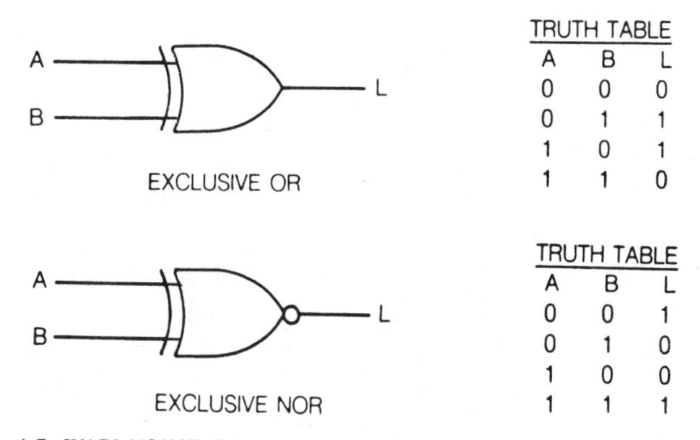

TRUTH TABLE

A	B	L
0	0	0
0	1	1
1	0	1
1	1	0

TRUTH TABLE

A	B	L
0	0	1
0	1	0
1	0	0
1	1	1

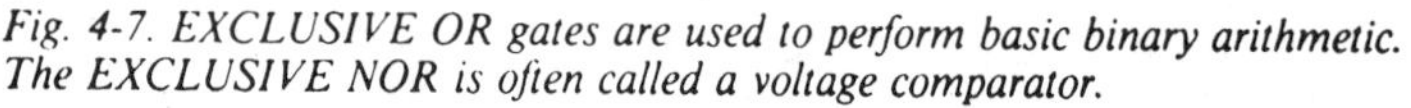

Fig. 4-7. EXCLUSIVE OR gates are used to perform basic binary arithmetic. The EXCLUSIVE NOR is often called a voltage comparator.

The truth table for the Exclusive Or gate shows that the only way to get a logic 1 output is to have the inputs at opposite logic levels. If the inputs are the same—that is, both zeros or both ones—the output is at a logic 0. The Exclusive OR gate is used in arithmetic and logic circuits. By comparison, the Exclusive NOR or Logic Comparator will produce an output only when the inputs are identical.

Both of the circuits in Fig. 4-7 are used in parity checks. A parity check is used to determine if a *word* [or combination of binary digits (bits)] is correct for a particular operation.

Suppose, for example, it is necessary to have an odd number of ones and zeros in order for the number to be correct. Using the Exclusive OR, the two zeros (an even number) and the two ones (even numbers) would show no parity but, the 0 and 1 and 1 and 0 inputs produce an output 1 indicating that parity exists. You should understand that any of the gates discussed so far, except the Inverter, may have as many as 16 inputs in a particular application. If, for example, there were 16 inputs to the Exclusive NOR, all of them would have to be at logic level 0 or logic level 1 in order to get a logic 1 output.

Timing Diagrams. In all of the gates discussed so far, the input levels were considered to be logic ones and zeros and the outputs were either at logic 1 or logic 0.

Another way to view logic gates is illustrated in Fig. 4-8. It is called a timing diagram. Instead of input logic levels, the three inputs to the NOR gate are pulse signals. The question is, how do you determine the output?

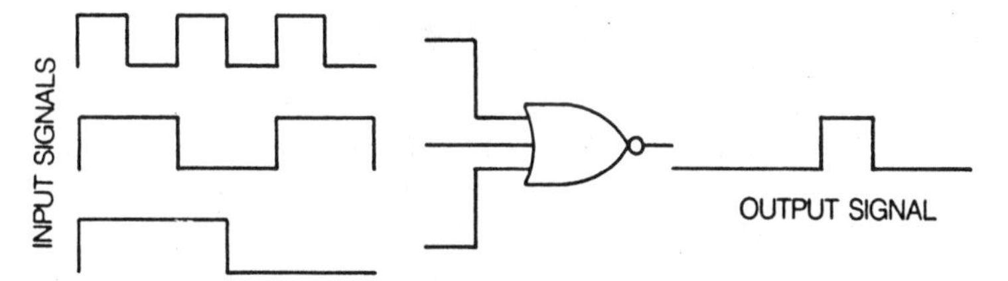

Fig. 4-8. Instead of combinations of ones and zeros, the input to a gate may be shown as signals. There is only one place in this illustration where all the inputs are at a logic. Flip flops are sometimes referred to as cross-coupled gates or data latches.

Remember, with the NOR it is only possible to get a logic 1 output if all the inputs are at logic 0. With the three input signals shown, that only occurs during the period when the three inputs are at a logic 0. At that time, the output is at a logic 1, as indicated by the output signal. There is really no new information here except that you have to remember that the input signals must all be in the right condition to produce the desired output.

FLIP FLOPS

Flip flops are bistable devices used extensively in logic systems. Traditionally, the outputs of these flip flops are marked Q and Q Not ($\overline{Q}$). The simplest possible flip flops are made with either NORs or NANDs as shown in Fig. 4-9. Some authors prefer to call these cross-coupled gates.

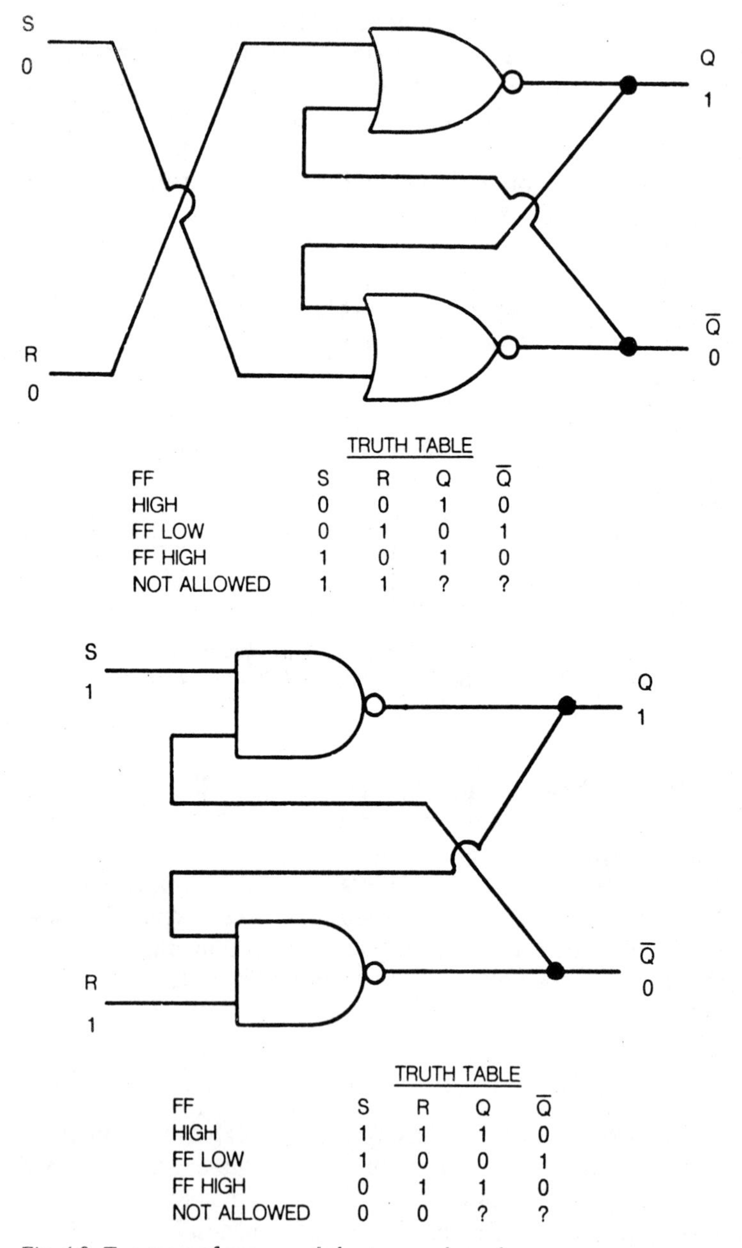

TRUTH TABLE

FF	S	R	Q	$\overline{Q}$
HIGH	0	0	1	0
FF LOW	0	1	0	1
FF HIGH	1	0	1	0
NOT ALLOWED	1	1	?	?

TRUTH TABLE

FF	S	R	Q	$\overline{Q}$
HIGH	1	1	1	0
FF LOW	1	0	0	1
FF HIGH	0	1	1	0
NOT ALLOWED	0	0	?	?

Fig. 4-9. Two types of cross-coupled gates are shown here.

Let's look at the NOR flip flop first. The input terminals are marked R and S for Reset and Set. So, this type of flip flop is sometimes called an R-S, or an S-R flip flop.

The truth table shows that there is a "Not Allowed" condition for the input to this flip flop. Both inputs must not be at logic 1 at the same time. If they are both at a logic 1 level, a "race" condition exists and the output cannot be determined.

If both inputs to the flip flop are at the logic 0, it is in a stable condition. The truth table shows this to be with a logic 1 at Q and a logic 0 at Q Not. However, this is an initial condition for the truth table. It could also be stable with the logic 1 at Q Not and the logic 0 at Q. The two possibilities are shown in the following table:

LOGIC LEVELS	NAME
$Q = 1, \overline{Q} = 0$	High
$Q = 0, \overline{Q} = 1$	Low

In order to get the flip flop to change its condition, a logic 1 must be delivered to one of the two input terminals. A logic 1 is delivered to the R while the S is held at logic 0. The output of the flip flop is Low, that is Q is at 0 and Q Not is at 1. If a logic 1 is delivered to S while the logic 0 remains at R, the flip flop will switch to a High condition.

If the flip flop is already in a Low or a High condition, it cannot be changed by the corresponding input signals. For example, when the flip flop is in a Low condition (Q = 0 and Q Not = 1), a logic 1 delivered to the reset cannot change it because it is already in a low condition.

The situation for the NAND flip flop is opposite. The Not Allowed condition for a NAND is two input zeros. A stable condition exists when both inputs are at logic 1. A logic 0 to either S or R can change the flip flop if it is not already in that condition. R-S flip flops are very often used as parts of a more complicated flip flop called the J-K type.

An Example of Flip Flop Switching. Figure 4-10 shows a NAND flip flop with input switching signals. In the first illustration, the NAND is in a stable High condition with two logic 1 inputs. In the second illustration, a logic 0 is delivered to the R input terminal. This switches the flip flop to a Low condition.

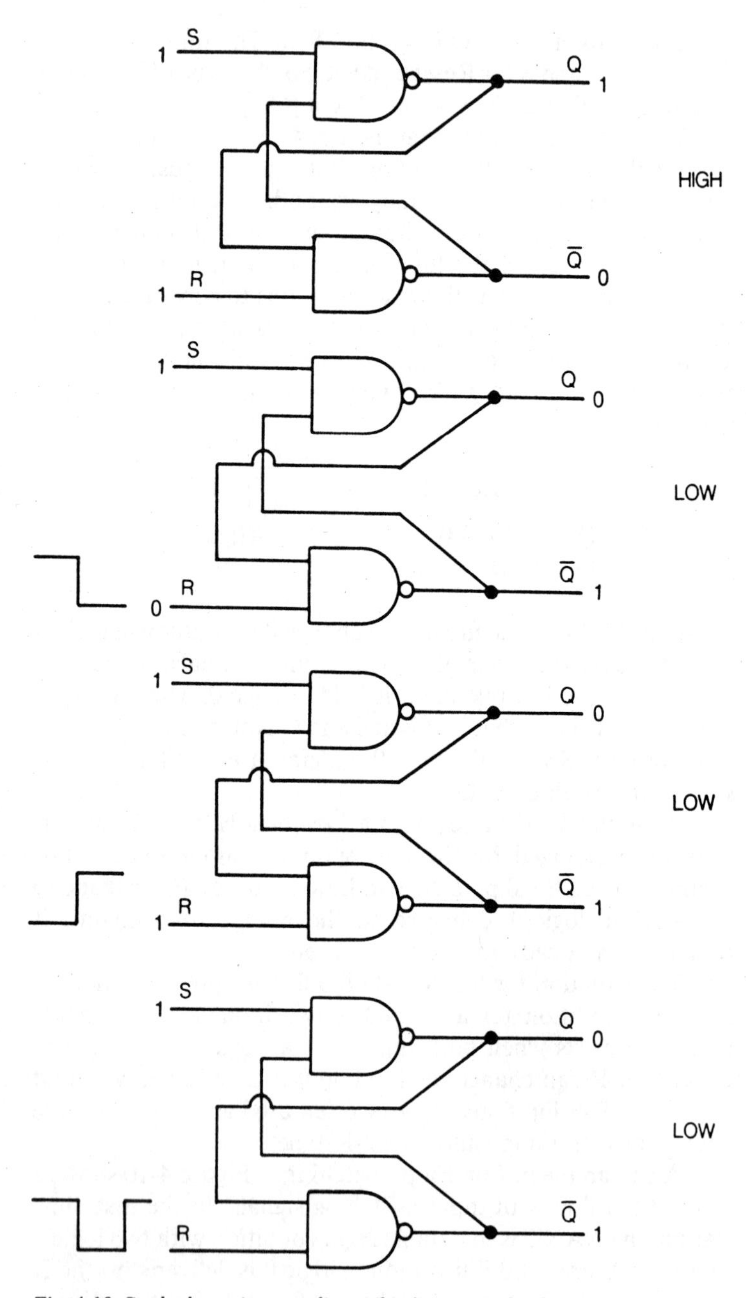

Fig. 4-10. Study the various conditions for the NAND flip flop in this illustration.

When the input is returned to a logic 1 as shown in the next illustration, the flip flop remains Low. The importance is that the flip flop "memorizes" the last input signal.

In the last illustration, a negative pulse, or logic 0, is delivered again to the reset terminal, but the flip flop state does not change because it is already Low. In order to change this flip flop back to the high state, the R must be maintained at logic 1 while a logic 0 is delivered to the S input terminal.

JK Flip Flops. The JK flip flop is more complicated than the RS type. It is sometimes called a master–slave flip flop because it consists of two flip flops ganged together in the same package. If you look at the symbol for the JK flip flop in Fig. 4-11, you will see that it contains J and K inputs and a reset and clear input that corresponds to the R and S of the internal RS flip flop.

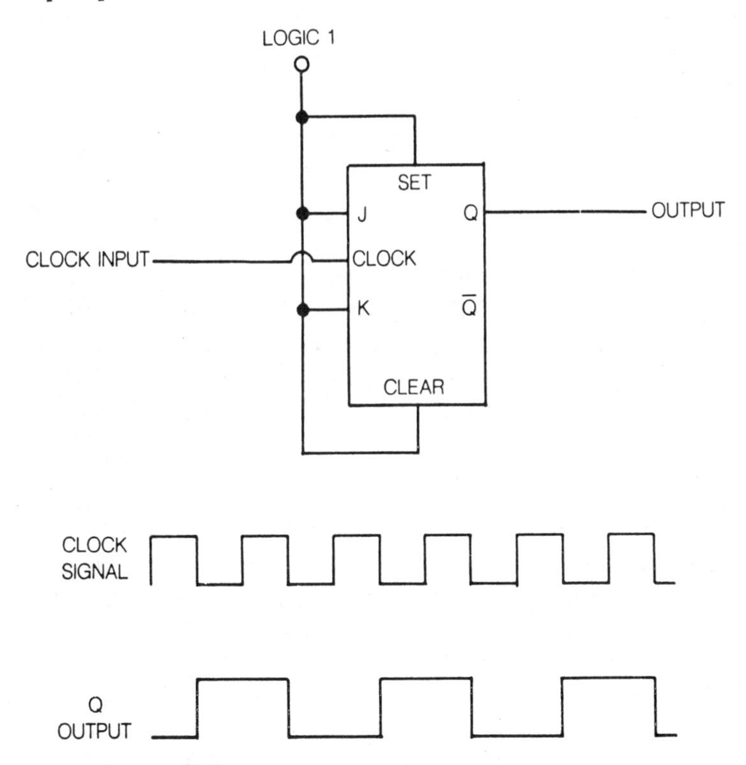

Fig. 4-11. The JK flip flop is wired to toggle. It divides the clock input signal by 2.

The J-K flip flop in Fig. 4-11 is wired to toggle. This simply means that it will change its output condition whenever the input clock signal goes from a 1 to a 0. This is presumed to be a TTL flip flop. If you are looking at a CMOS flip flop, it will change state on the leading edge rather than on the trailing edge of the clock signal.

If you look at the timing diagram, you will see that there is a change in the output at Q each time the clock goes from one to zero. Since that only occurs at the end of each cycle, the Q output has a frequency that is exactly one half of the input frequency. Therefore, the toggled flip flop is sometimes referred to as a divide-by-two circuit.

Remember that the JK flip flop can be used for any application where flip flops are desired, but this particular application is especially important because it is used in counters and in divide-by circuits.

Consider now two toggle flip flops connected together as shown in Fig. 4-12. Here, the input signal is on the right side. From Fig. 4-11, you know that the set, JK, and clear inputs must all be at logic 1 for the flip flop to toggle. The first flip flop (that is, the one on the right) divides the clock signal by 2 and the second flip flop (that is, the one on the left) divides the signal by two again.

The overall result is that the output at the most significant bit (MSB) is one fourth the frequency of the clock input and the least significant bit (LSB) is one half the clock input signal. With a clock input, the input logic levels of the MSB and the LSB are shown in the table. If you look at it carefully you can see that the circuit is counting in binary.

After the count has reached decimal 3 (binary 11), it starts over. The circuit in Fig. 4-12 is called a ripple counter, or ripple through counter. The count is made by converting the first flip flop, then the second one. If there is a longer series of flip flops for counting more digits, it takes a certain amount of time for this ripple through counter to change all of the flip flops—one at a time—to reflect a particular count. So, its disadvantage is that it is relatively slow, although it has a good power supply advantage. With each flip flop changing state one at a time, the power supply does not have to deliver a large amount of power to make a count.

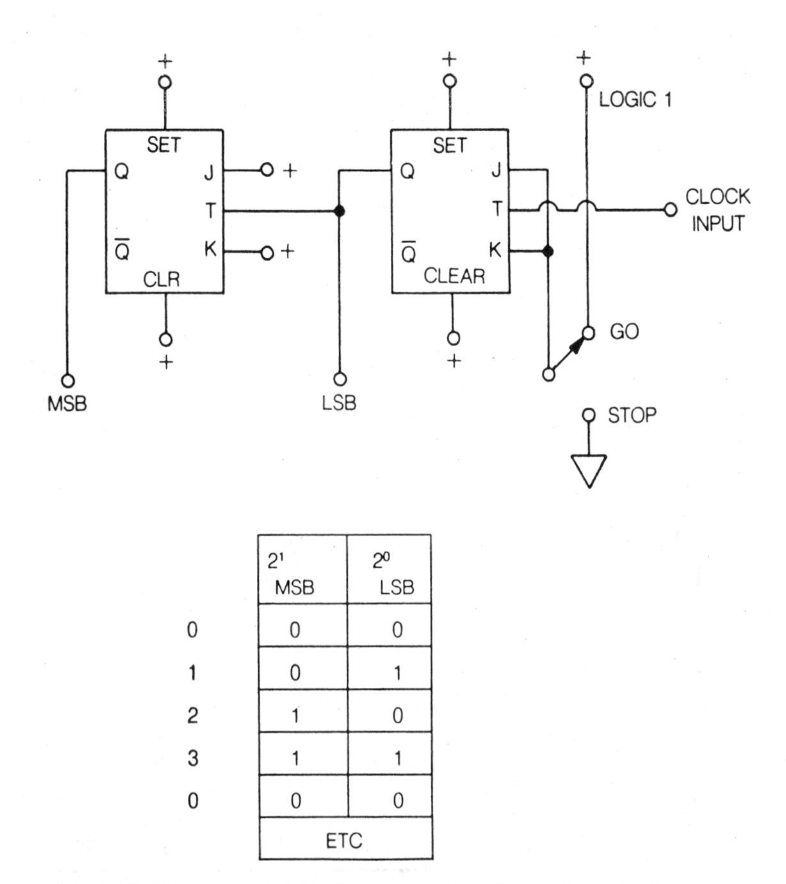

	2^1 MSB	2^0 LSB
0	0	0
1	0	1
2	1	0
3	1	1
0	0	0
	ETC	

Fig. 4-12. The two toggled flip flops are combined to make a simple counter.

Now consider the synchronous counter illustrated in Fig. 4-13. Observe that the clock signal is delivered to all of the JK flip flops at the same time. The J and the K of the first flip flop on the right is wired to logic 1 so that it can toggle. Toggling of the other flip flops occurs when the output from the Q of the first flip flop is at a logic 1. This type of counter is very fast. If you want to add more flip flops to the one in Fig. 4-12, you can just follow the same pattern.

In order to add another flip flop to the one in Fig. 4-13, you need to employ external gates. In the illustration shown in Fig. 4-13, flip flop number 3 is added by taking the LSB and MSB of the upper counter into an AND circuit. When those two bits are

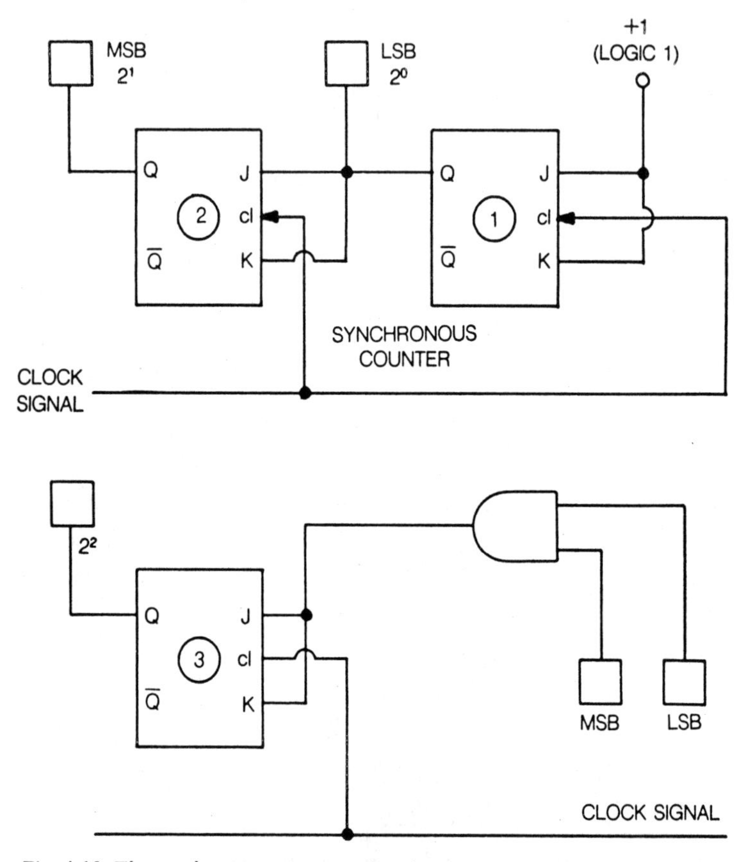

Fig. 4-13. The synchronous counter makes the same count as the ripple counter in Fig. 4-12. To add an additional stage to this counter, employ an AND gate as shown for flip flop no. 3.

at a logic 1, there will be a logic 1 delivered to the J and K of flip flop number 3.

The powers of 2 shown in Figs. 4-12 and 4-13 simply show the positions of the binary digits to make the complete binary number. So, the positions of the digits are identified either by a power of 2 or by the value of that power of 2, as:

8	4	2	1
2^3	2^2	2^1	2_0
1	0	0	1

Example:

$$1 \times 1 = 1$$
$$2 \times 0 = 0$$
$$4 \times 0 = 0$$
$$8 \times 1 = 8$$
$$\text{Total} \quad 9$$

Note that in both the synchronous and ripple-through counters, the first JK flip flop must have the J and K at logic 1 in order to count.

Jamming. Counters can be stopped or reset (jammed) by delivering a logic 0 to the J and K or to the reset terminal.

The procedure for stopping or resetting the count is illustrated in Fig. 4-14. Here we have four digits set up by four flip flops in a counter. The number represented is binary nine. When that number occurs, two ones will be input to the NAND. This results in a 0 output, which, when delivered to the JK input, stops the first flip flop from counting. If you stop the first flip flop from counting, then all the other flip flops will stop counting also.

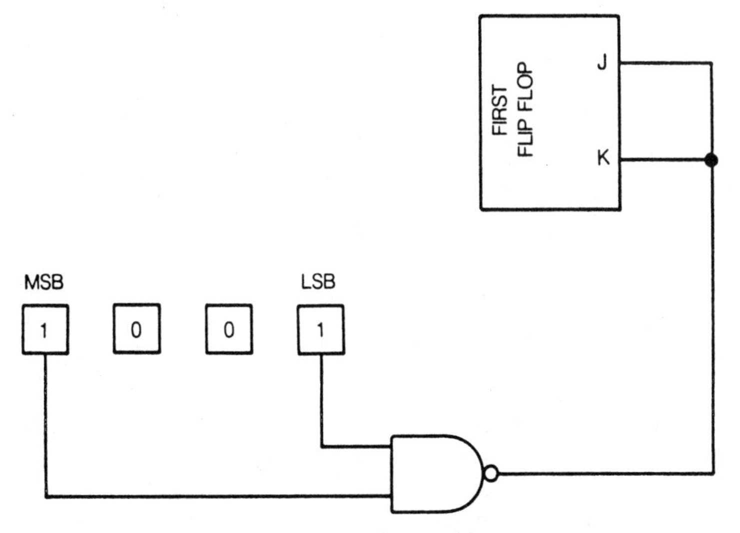

Fig. 4-14. The principle of jamming is illustrated here.

If the output of the NAND had been delivered to the reset, all the digits would have been returned to 0 on the next transition after 1001. In other words, the count would start over again at 0.

Flip flops can be made to count down instead of up by changing the Q connections to Q not connections in either of the flip flops. Jamming of that type of countdown circuit is accomplished the same way, as shown in Fig. 4-14.

To count up or down, it is customary to use a binary up-down counter available in all logic families. These counters consist of a large number of internal flip flops that can be programmed, or jammed, using the proper input voltages on their terminals.

The Data Flip Flop. An important feature of flip flops is their ability to "remember" the last input signal. This characteristic is used in Random Access Memories (RAMs) by using a special input with an R-S flip flop. It is shown in Fig. 4-15. RAMs are discussed in the next section.

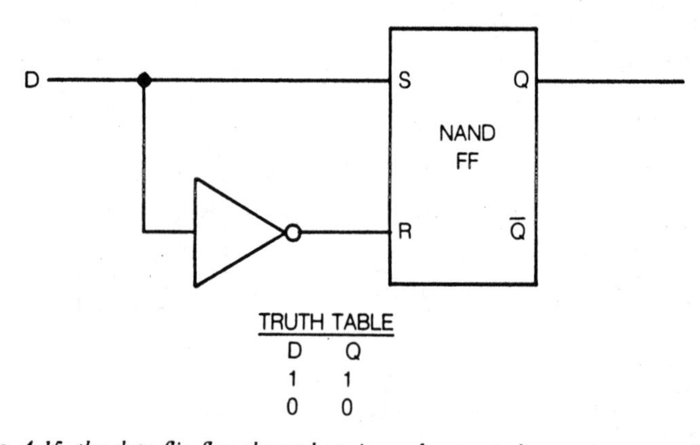

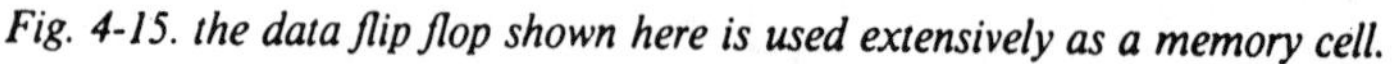
Fig. 4-15. the data flip flop shown here is used extensively as a memory cell.

This D flip flop, or data flip flop is set Low or High, depending on the test data input. For example, if the last data input was Low, there will be a Low on Q and a High on Q Not.

Data flip flops are very often used as memory cells in large integrated circuit memory. They are set High or Low, depending on whether a logic 1 or logic 0 is to be stored.

MICROPROCESSORS

Modern digital systems make extensive use of memories. Two very important kinds of memory are RAM and ROM. RAM (Random Access Memory) can be used to store combinations of logic ones and zeros in a fixed pattern called a word. In some microprocessor systems, a word consists of eight individual bits of data. Bits are *b*inary dig*its* made up of either ones or zeros. An 8-bit word is called a byte. A 16-bit word (two bytes) is common in many modern microprocessor systems. It has 16 binary digits, called two bytes. A nibble is four bits.

The digits are stored in RAM addresses. An address is itself a binary number. Putting the information into or taking it out of RAM involves selecting the address with a binary number, and then applying the combinations of ones and zeros for storing the word. Also, the address permits the word to be retrieved.

The name Random Access Memory is not an accurate description of its operation. Most memories used in electronics can be accessed at random by using the right address. ROM (Read Only Memory) also stores information, but the information is put in during manufacture or during a special process in which the complete memory is loaded. Except for the method of putting the memory in and the fact that you cannot write information in ROM, it can be accessed as easily as RAM.

You would use ROM in a system, for example, to store the value of *pi* or to store the step-by-step procedure for solving a basic equation. So, think of ROM as being a permanent memory and RAM as being one in which information can move in and out of freely. Since you can go to a specific location in either RAM or ROM to get a word or combination of words from memory, they are both randomly accessible. RAMs are also sometimes called read-write (R/W) memories.

A microprocessor is a system that is designed to use memories. A typical microprocessor system is shown in Fig. 4-16. All microprocessors require a clock signal to step through their range of operations. Considerable emphasis is placed in modern microprocessor systems on how fast they can perform certain applications such as arithmetic and logic problems. So, the clocks have frequencies that are usually between two megahertz and twenty megahertz. There are some specialized applications outside this range of frequencies.

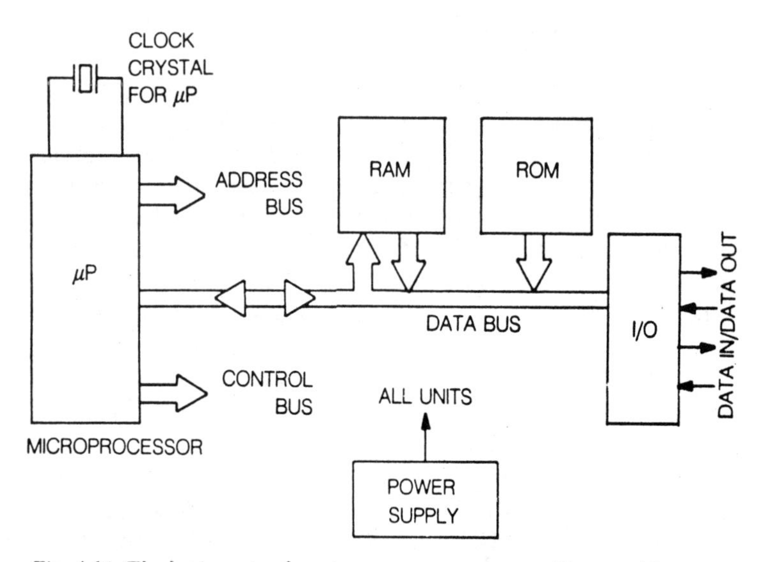

Fig. 4-16. The basic parts of a microprocessor system are illustrated here.

The microprocessor operates by using three very important buses. A bus is a combination of parallel conductors. The data bus is used to deliver the combinations of ones and zeros necessary to make up a word. If you are going to store a word in RAM or take a word out of RAM, the word would be delivered on a data bus. Also, if the microprocessor gets information from ROM or from an outside (peripheral) device like a keyboard, the information arrives on the data bus.

The address bus tells what location the word is going to go to or where it is to come from. The control bus turns the necessary memories or other circuits on and off with special coded signals.

A typical operation of a microprocessor might be to add two numbers. The two numbers are stored in RAM, or sometimes in ROM. The microprocessor gets these numbers from memory and stores them in special short-term memories called registers or accumulators—depending upon the manufacturer of the microprocessor. Once the numbers have been retrieved and stored, the clock signal steps the microprocessor through the necessary procedure to add the numbers and store the result. The result may be put into RAM. The microprocessor also interfaces with the "outside world." Instead of getting the numbers to be added from RAM, they can be obtained from a keyboard that delivers the signal through the input/output (I/O) peripheral interface. Likewise, instead of storing the result in RAM, the number may be delivered to the peripheral interface and from that point to a cathode ray tube usually (called the monitor) in microprocessor systems. Microprocessors can perform arithmetic and logic functions, and are often used in computers for that purpose.

The difference between a simple microprocessor system and a large computer is primarily in the amount of memory and in the methods of delivering and retrieving data.

Computers usually are rated by the amount of memory they have. Memory is measured in megabytes (millions of bytes). In industrial electronic systems, computers are used to control machinery, store procedures, control manufacturing, programming robots and a wide variety of other applications.

SUMMARY

In industrial electronics, the use of digital circuitry and microprocessors has become extremely important. This includes computers, which employ microprocessors.

You also hear much about CAD (Computer Aided Design) and CAM (Computer Aided Manufacturing). These systems make use of computers to perform their design and manufacturing functions.

Microprocessors and computers are used extensively in robotics and other assembly-line operations. Digital circuits are part of every microprocessor and computer system. Microprocessors are also used in numerical control systems and in other automated systems. You need to know how these systems work in order to effectively troubleshoot them when they are not working.

It is not possible to cover all the technology involved in digital and microprocessor systems. This chapter covers the essentials that you are expected to know as an industrial technician at an entry Journeyman level. If you find that some of the subjects covered in this chapter are not familiar to you, I suggest that you visit an electronics distributor. There are some very good books that cover all of these subjects in depth.

You may be interested in taking the Journeyman CET Test for Industrial Electronics. If so, it would be a good idea to examine the appendix in this book.

SELF TEST

1. Information in a microprocessor is transported from a register to memory by way of the:
   (A) control bus.
   (B) data bus.

2. RAMs are often made with:
   (A) fusible links.
   (B) data flip flops.

3. Every microprocessor uses:
   (A) a clock signal.
   (B) an accumulator.

4. The only way to get a zero out of a NAND gate is to have both inputs at a:
   (A) logic 0 level.
   (B) logic 1 level.

5. Another name for an Exclusive NOR gate is:
   (A) inclusive OR gate.
   (B) logic comparator.

6. A microprocessor that is not in an integrated circuit package, but rather is made of individual integrated circuits is called:
   (A) mainframe.
   (B) motherboard.
   (C) Neither choice is correct.

7. A single input is used to feed a number of different lines. It is called a:
   (A) multiplexer.
   (B) demultiplexer.

8. Is the following Boolean algebra solution correct?

$$A\ (A + B) + AB = A + AB$$

   (A) Yes.
   (B) No.

9. According to DeMorgans theorem, $A + B$ equals:
   (A) A *or* B
   (B) $A - B$.

10. The disadvantage of a ten-digit synchronous counter is that:
   (A) it is slower than an asynchronous counter.
   (B) it places a heavy demand on the power supply each time a count is made.

ANSWERS TO SELF TEST

1. (B)

2. (B) Fusible links are used in ROMs.

3. (A) One of the first steps in troubleshooting a microprocessor system is to make sure the clock signal is OK.

4. (B)

5. (B)

6. (C) The correct answer would be bit slice. One advantage of a bit slice is its high speed.

7. (B)

8. (A) SOLUTION:

$$
\begin{aligned}
A(A + B) + AB &= AA + AB + AB \\
&= AA + A(B + B) \\
&= A + AB
\end{aligned}
$$

9. (A)

10. (B) A synchronous counter is faster than an asynchronous counter.

5

Control of Power

BY FAR THE most frequently encountered power sources are single phase. The reason for this is that even though the major power system is three phase, power for electronic equipment is often taken across one of the three phases. So, looking out from the electronic equipment you would see a single-phase source.

As an electronics technician you are expected to know some basics of three-phase theory and some basic three-phase rectifier circuits. They are included in this chapter for review.

Two-phase power is not used extensively in today's systems. An exception is in capacitor start motors. They behave as though connected to two-phase power during startup. After a predetermined speed is reached, the capacitor that sets up the two-phase operation is switched out. Rectifier and unregulated power supplies are important subjects in this chapter.

CHAPTER OBJECTIVES

After you have studied this chapter you will be able to answer these questions:

- What is a rotary converter used for?

- What is the difference between an inverter and a converter?
- How is a transformer used to match impedances?
- What is the difference between a self-saturating and a ferro-resonant transformer and how are they used?
- How is a conjugate match obtained?

POWER CONVERSION

All electronic systems require some kind of power source. It can be as simple as a battery, or as complicated as a three-phase power system.

Conversion from one form of power to another is also an important part of most electronic systems. For example, if dc is needed for operating electronic equipment and it must be obtained from the power line, then a power supply rectifier is needed. There are four basic kinds of power conversion illustrated in Fig. 5-1.

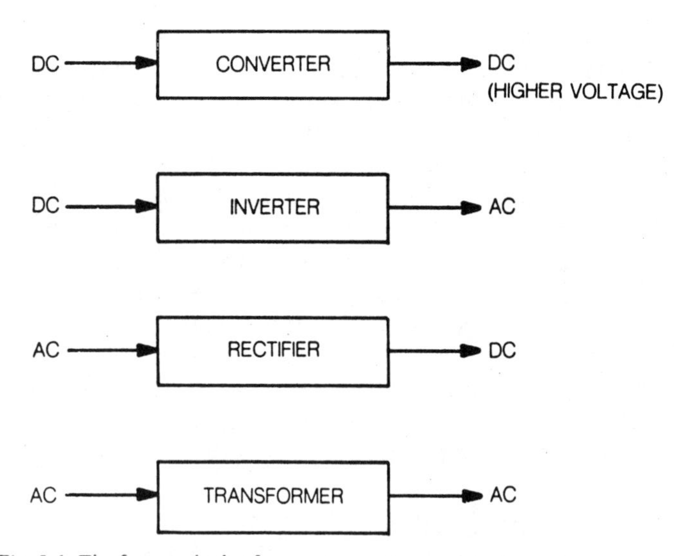

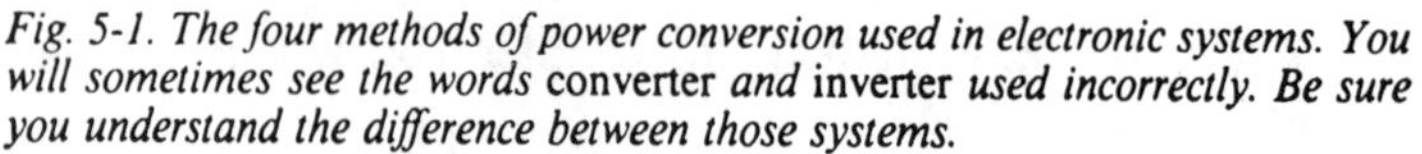

Fig. 5-1. The four methods of power conversion used in electronic systems. You will sometimes see the words converter *and* inverter *used incorrectly. Be sure you understand the difference between those systems.*

Kinds of Power Conversion. To convert from one dc voltage to a higher dc voltage, a converter is used. To go from a higher voltage to a lower voltage, it is customary to use a voltage divider.

One way of converting from a lower dc voltage to a higher dc voltage is to use a rotary converter. This is simply a motor-generator set with the motor and generator on the same shaft. The lower dc voltage to be converted drives the dc motor and the higher dc voltage is taken from the generator output.

An inverter is used to convert from dc to ac. It is represented in Fig. 5-1. The usual approach is to use the dc to operate some form of oscillator. The output of the oscillator becomes the ac. The ac voltage can be changed from one level to another by using a transformer. Some converters use inverter circuitry. They first change the dc to ac, step the voltage up through a transformer and then rectify the higher voltage to get the desired dc level. This concept is shown in Fig. 5-2.

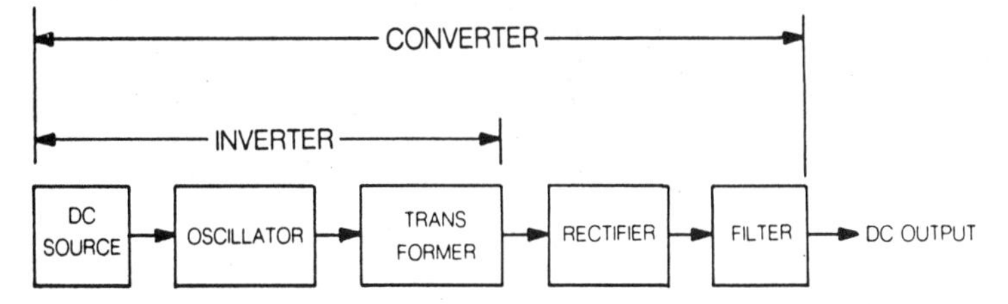

Fig. 5-2. In this illustration you see that an inverter is always part of a converter circuit.

Rotary inverters are also available. They convert dc to ac by having a dc motor drive an ac generator. Traditionally the dc motor and ac generator are on the same shaft. As shown in Fig. 5-2, an electronic converter has an inverter in its circuitry.

Inverters and converters are usually treated separately. A rectifier converts ac to dc. There are a number of rectifier circuits used to do this. They will be discussed in this chapter.

The last conversion illustrated in Fig. 5-1 is from ac to ac. This is easily accomplished with a transformer. Conversion to a lower ac voltage could be accomplished with a voltage divider, but transformers are more efficient.

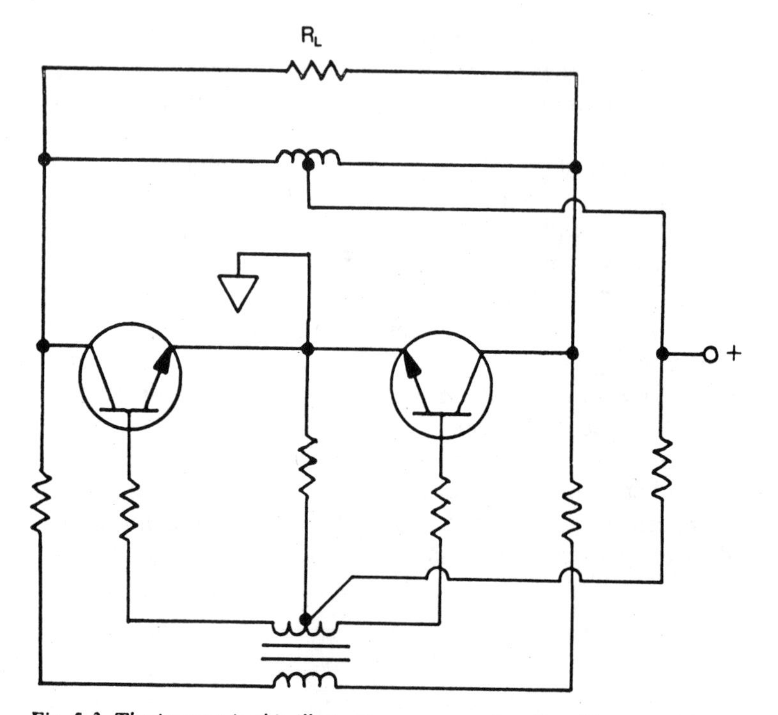

Fig. 5-3. The inverter in this illustration uses a push-pull oscillator.

The Inverter/Converter. Figure 5-3 shows an inverter circuit. Basically, every oscillator is an inverter because an oscillator can be defined as a circuit that converts dc to ac. The oscillator shown in this inverter is a simple push-pull type. Inverters are used extensively in portable equipment that must be battery operated and an ac voltage is desired for operating some types of ac appliances.

Oscillators are also used in power supply switching regulators. In that system, an oscillator—operating on dc—produces a pulse output. The width of the pulse is controlled to adjust the output power.

Figure 5-4 shows a block diagram of an oscillator. If you can identify these sections in an oscillator, you will find it easier to troubleshoot the circuit.

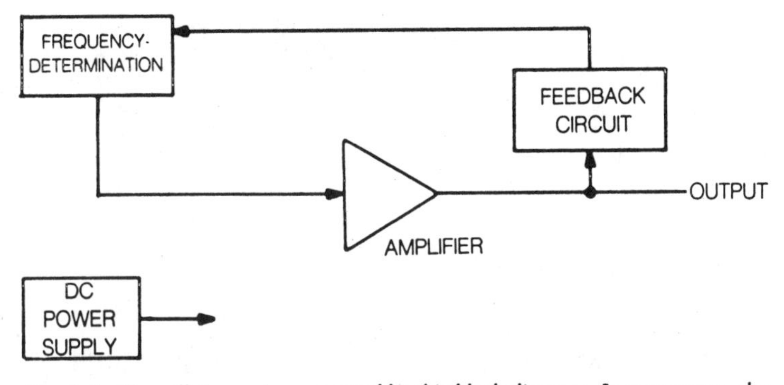

Fig. 5-4. All oscillators are represented by this block diagram. In some cases, the feedback and frequency determination circuits are combined into one circuit.

REGULATED POWER SUPPLIES

Having looked at the various types of power supplies in general, we will now take a look at the most popular and most important power supply circuit in electronic systems—the regulated power supply. Figure 5-5 shows a block diagram of a typical analog regulated supply.

Each section of the supply in Fig. 5-5 will be discussed in detail. Some of these sections, such as the transformer and the rectifier, are also used in other types of regulated and non-regulated supplies.

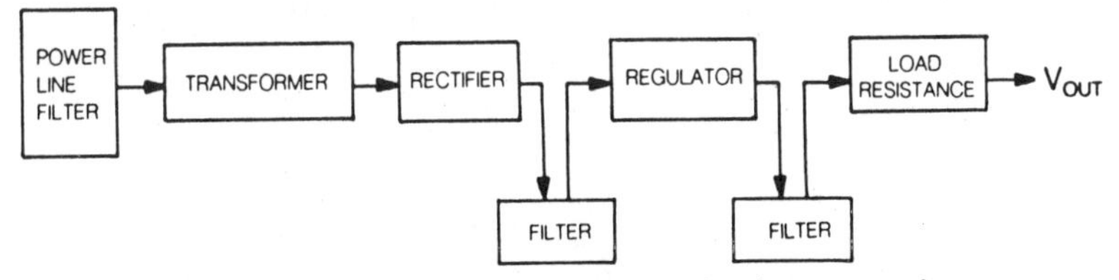

Fig. 5-5. This is a block diagram of a simple regulated power supply.

The Transformer. Figure 5-6 shows the symbol for a typical power transformer. The dots on the transformer symbol indicate the points that are in phase. It is important to know the phase relationships between transformers in certain applications such as inverter—oscillator circuits. The broken line between the primary and secondary windings represents a Fara-

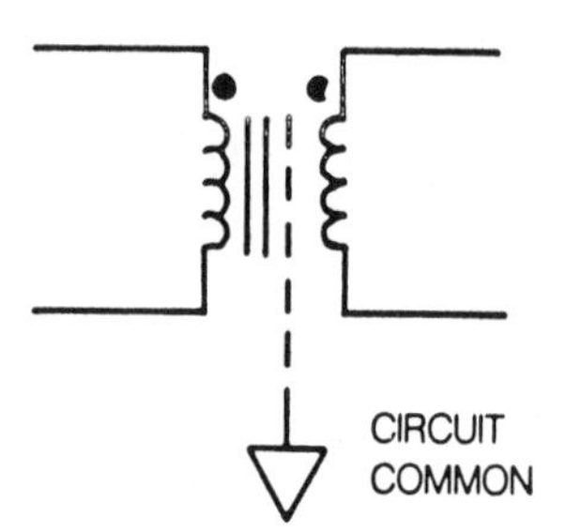

Fig. 5-6. Two important points are illustrated in this transformer schematic. The dot notation shows the points on the transformer primary and secondary that are in phase. The broken line shows how a Faraday screen is connected in a power transformers.

day shield. The Faraday shield is an electrostatic shield. It prevents coupling of transient voltages from the primary to the secondary through the primary-to-secondary capacitance. Remember that the primary and secondary of a transformer are conductors and they are separated by an insulator. Any time you have two conductors separated by an insulator, a capacitance is present. That capacitance offers an easy path for high-frequency transient voltages such as noise spikes. The Faraday shield prevents those voltages from being coupled through the transformer.

The following equations are standard for relating the turns ratio to voltage relationships, current relationships, and impedance relationships between the primary and secondary.

$$\text{TURNS RATIO} = \frac{N_p}{N_s}$$

$$\frac{N_p}{N_s} = \frac{V_p}{V_s}$$

$$\frac{N_p}{N_s} = \frac{I_s}{I_p}$$

Note: This equation is important for impedance matching. This is not a main function of power transformers:

$$\frac{N_p}{N_s} = \sqrt{\frac{Z_p}{Z_s}}$$

From which:

$$Z_p = \left(\frac{N_p}{N_s}\right)^2 Z_s$$

where Z_p is the reflected impedance.

Sample calculation:

The turns ratio of a certain transformer is 1/10. How much impedance is reflected into the primary if the load impedance is 1000 ohms?

Answer: $Z_p = (1/10)^2 \times 1000 = 10$ Ohms

Figure 5-7 shows how the turns ratio can be used to change the primary to secondary voltage. A question sometimes asked in entry-level industrial electronics CET tests is given here:

For maximum possible output voltage, what setting of the transformer in Fig. 5-7 is required for the switch?

Answer: The highest secondary voltage will occur when the switch is in position 3. In that position there is the greatest ratio of secondary-to-primary turns and therefore the greatest ratio of secondary-to-primary voltage.

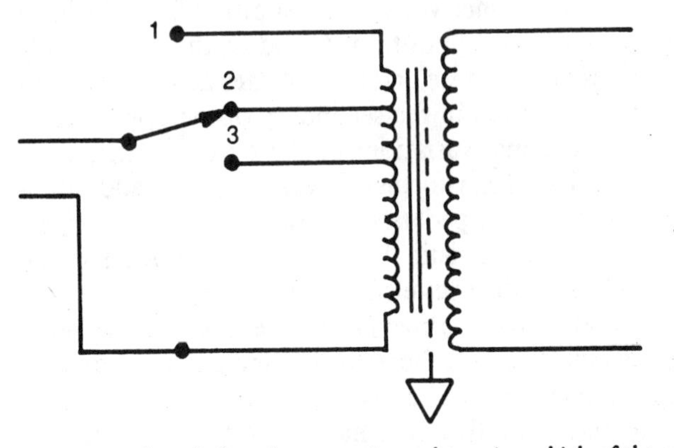

Fig. 5-7. Use your knowledge of turns ratio to determine which of the switch positions will produce the highest output voltage.

Transformers are sometimes used to compensate for minor changes in line voltage. An example is shown in Fig. 5-8. The broken line between the switch contacts indicates that they are

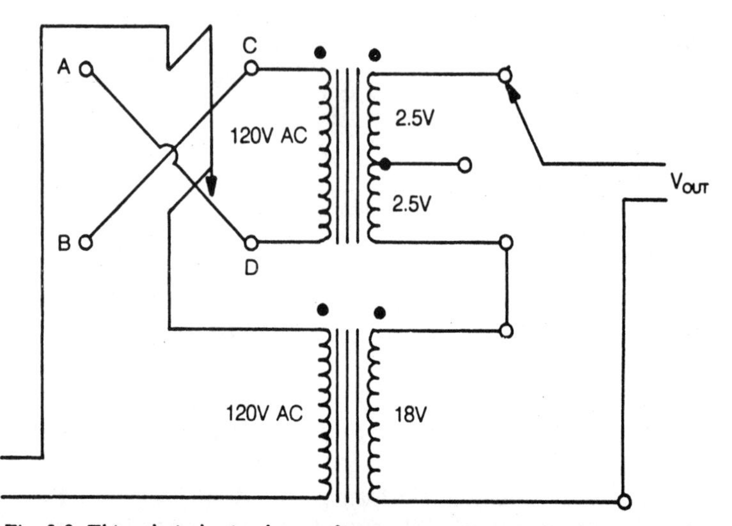

Fig. 5-8. This relatively simple transformer connection permits the output voltage to be adjusted to compensate for input ac line voltage changes.

mechanically connected. When they are in the position for A and B, the upper transformer has a phase that is opposite the phase of the primary of a lower transformer. That means that the upper transformer voltage will subtract from the lower one. When the switch is in positions C and D, the two primaries and secondaries are in phase and the voltages across the secondaries will add. The switch in the secondary of the upper transformer provides additional variations of the output voltages.

Specialized Transformer Operations. In addition to voltage, current, and impedance ratios, there are some specialized transformer characteristic that you should know. They are used extensively in industrial electronics systems.

Regulating transformers are sometimes used as power supply preregulators. In other words they are used ahead of the electronic regulators. The two types are shown in Fig. 5-9.

The self-saturating type has a core that becomes saturated before the peak of the primary current is reached. Once the core is saturated, there can be no further change in the secondary output voltage. Therefore, the saturation limits the output voltage to a specific value. If the primary is operated at a point below saturation, no regulation can take place.

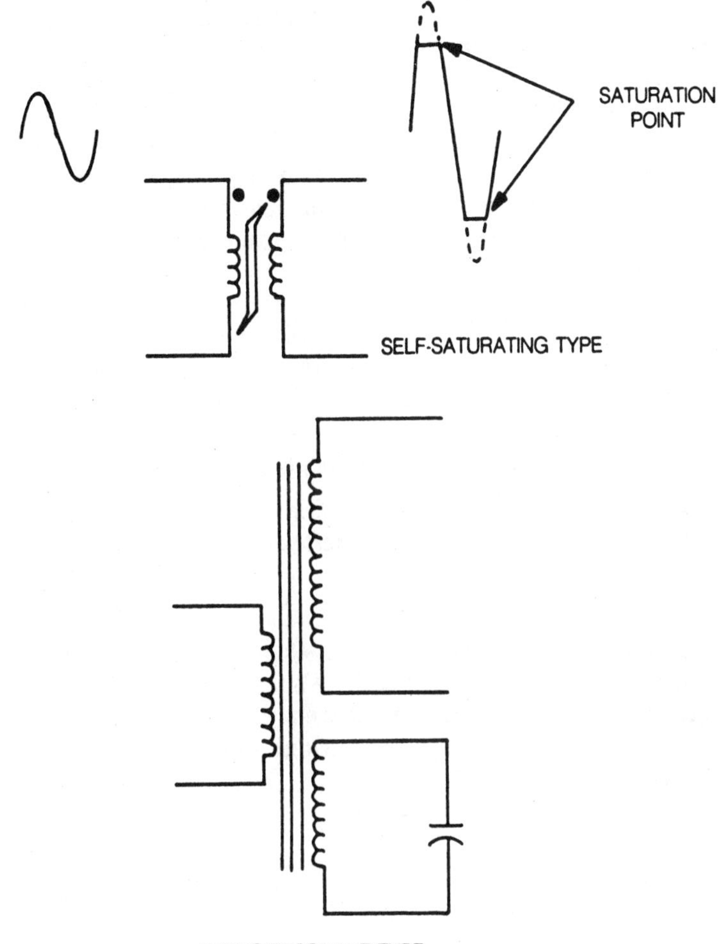

Fig. 5-9. The self-saturating transformer goes into saturation on the peaks of the input ac voltage. Once the peak value is reached, the output can no longer change. Therefore, the output voltage waveform is squared at the top and bottom. The ferro-resonant transformer uses a resonant circuit (comprised of L and C) to drive the transformer core into saturation. Saturation occurs when the input voltage nears its peak value.

The ferro-resonant transformer has a resonant circuit made with a separate winding and a capacitor. In an alternate design, the capacitor is across part of the secondary winding. This tuned circuit is resonant at the primary frequency.

When ac current flows in the primary, the changing flux produces a *flywheel*, or back-and-forth, current in the LC circuit. When that current reaches its peak value, it drives the transformer into saturation.

Naturally, this circuit will not work if the capacitor is not present. Anytime you are troubleshooting a circuit that contains this type of transformer you should make it a point to check the capacitor if it is connected externally to the transformer.

The Faraday shield is usually connected to a braided wire that comes out from the bottom of the transformer. This braided wire must make a good connection to the common or ground of the electronic circuit in order for the Faraday shield to work properly.

Maximum Power Transfer Theorem.

> *The maximum possible power from a dc source occurs when the load resistance is numerically equal to the internal resistance of that source.*

The maximum power transfer theorem is very important because it permits the maximum possible power to be delivered to a load regardless of efficiency. Also, it shows the tradeoff between maximum power and efficiency. The best way to understand this concept is to start by using a simple dc circuit. (See Fig. 5-10).

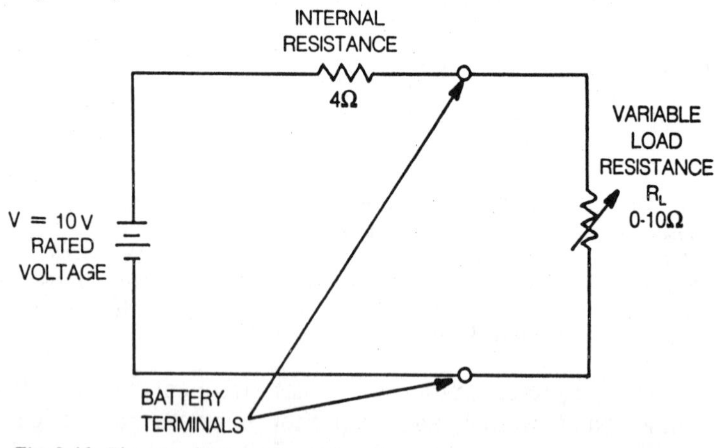

Fig. 5-10. This simple circuit is used to illustrate maximum power transfer and efficiency.

A 10-volt battery with an internal resistance of 4 ohms is connected to a variable load resistance that can be adjusted from 0 to 10 ohms. As that resistor is varied through its range of values, its power dissipation also varies.

Figure 5-11 shows a graph of the power dissipated by R_L versus the value of R_L. Observe that the curve goes through a maximum value of 6 watts at $R_L = 4$ ohms. That is the value of the internal resistance of the battery.

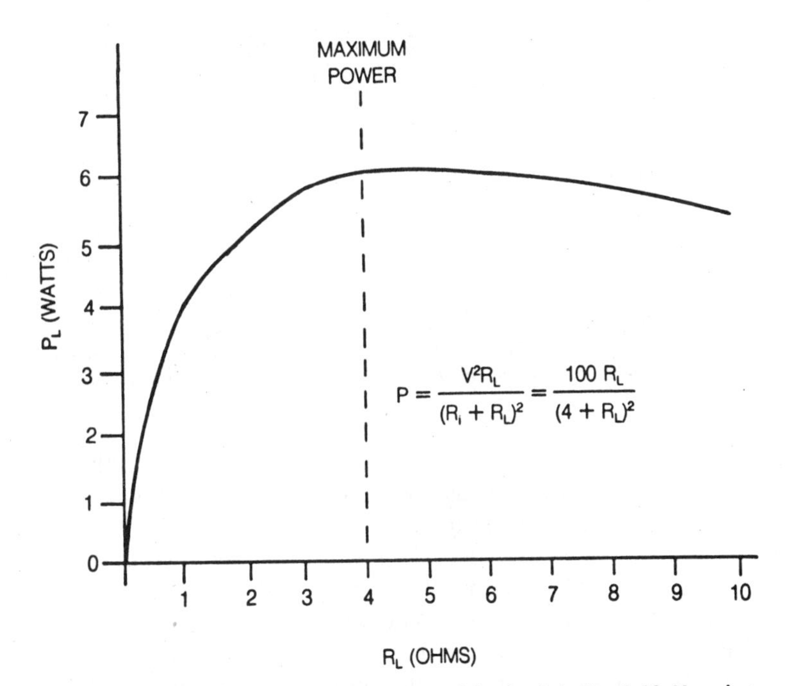

Fig. 5-11. The curve shows the output power of the circuit in Fig. 5-10. Note that the power goes through a maximum value when the load resistance is equal to the internal resistance of the power supply.

You might think, then, that all systems are operated under maximum power conditions. That is not the case because there is an important tradeoff between power and efficiency. The efficiency of a dc power source like the one in Fig. 5-10, is only 50 percent when the maximum power is being dissipated in the load resistance. This is also illustrated in Fig. 5-12.

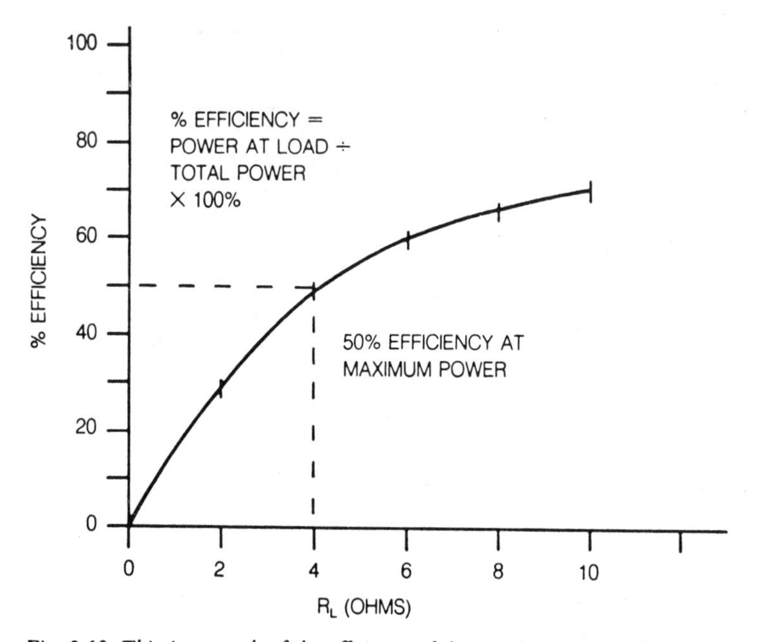

Fig. 5-12. This is a graph of the efficiency of the simple power supply in Fig. 5-10. Note that the internal resistance value for RL, the efficiency is only 50 percent.

Observe that at $R_L = 4$ ohms, the efficiency is only 50 percent. Note also that as the value of R_L increases above 4 ohms, the efficiency continues to increase.

Suppose you were designing a flashlight that used a single flashlight cell and a lamp. The maximum possible brightness from the lamp would occur under maximum power conditions. However, the life of the battery would seriously be shortened by that type of operation. On the other hand, if you make the efficiency too high, and the drain on the battery too low, you will not get enough light to make the flashlight useful.

This is one of the important tradeoffs that designers must face. As a technician you probably will not be doing design work—at least in the early stages of your work. But, you should understand the reasoning behind some designs in order to understand how the circuit works and what some of the parameters are based upon.

Maximum Power in AC Circuits. So far we have discussed the maximum power transfer theorem only as it applies to dc. This theorem has to be modified to work for ac circuits. To understand why, consider the ac generator shown in Fig. 5-13. Here the ac generator has an internal impedance consisting of an inductance and resistor in series. This would be a typical internal impedance for an ac electromechanical generator. There would also be some distributed capacitance, but the capacitive reactance and inductive reactance subtract. So, in this example, the equivalent circuit is simply the inductor and resistor in series.

Here is an important question: What would you connect across the output terminals of the generator in Fig. 5-13 in order to get maximum power transferred to the load?

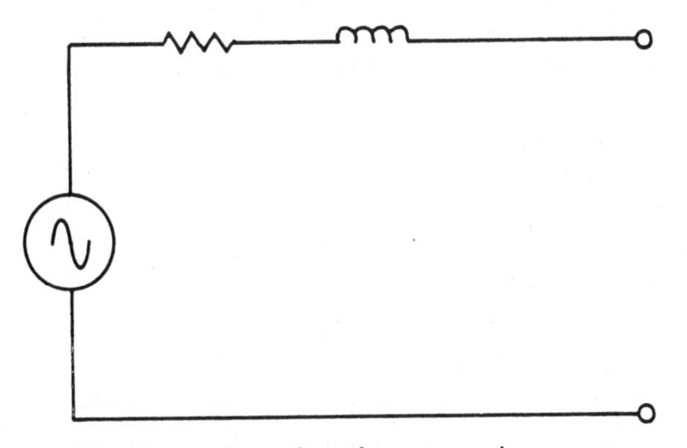

Fig. 5-13. A typical internal impedance for an ac supply.

Obviously, placing a resistor across the output terminals will not do it. The reason is that there is a phase angle between current and voltage caused by the reactance. Since the current and voltage do not peak at the same instant, it is not possible to get maximum power out of the ac generator with a resistive load.

The two illustrations in Fig. 5-14 show the types of load impedance that must be connected to get maximum power. In each case, the load impedance is called the *conjugate* of the internal impedance of the generator. So, the conjugate of $R + jX_L$ is $R - jX_C$. Likewise, the conjugate of $R - jX_C$ is $R + jX_L$.

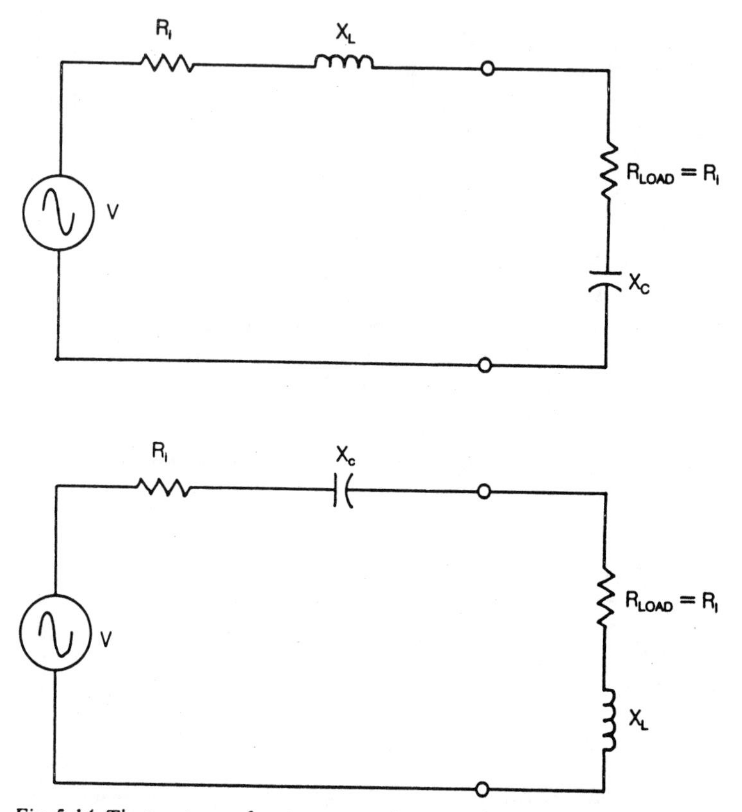

Fig. 5-14. The two types of conjugate matches. Note that the conjugate match is designed so that the load resistance is equal to the internal resistance, and the reactances cancel. The components of the impedance triangles for a conjugate match are also shown.

The impedance and power triangles for ac circuits are shown in Fig. 5-15. Only an $R - jX_C$ load impedance will be considered here, but with opposite signs the same thing applies to the circuit with a load impedance of $R + jX_L$.

Voltage and Impedance Triangles. Figure 5-16 shows the legs of the impedance triangles for the internal impedance and load impedance applicable to the circuit with an internal impedance of $R + jX_L$. This is a maximum power load impedance. Observe that in this case the X_L and the X_C cancel and the resulting load impedance is R_L. When that value of R_L is equal

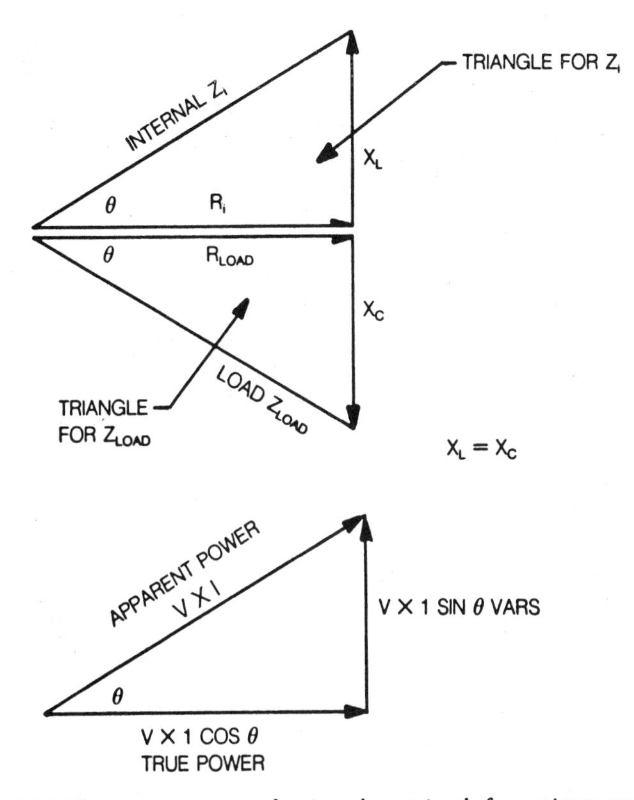

Fig. 5-15. The component parts of an impedance triangle for a cojugate match.

to the internal resistance R_i, maximum power transfer will be obtainable.

Using the same reasoning, the load impedance of the circuit with an internal impedance of $R - jX_C$ is the conjugate of that internal impedance when maximum power is to be transferred. The same illustration shown in Fig. 5-16 would apply.

Figure 5-15 also shows a typical power triangle. Traditionally, this triangle is shown in the first quadrant, that is, with a positive angle of θ between 0 and 90 degrees. The phase angle between the voltage and current is θ, as it is in the impedance triangles of Fig. 5-15.

The *apparent power*, or V × I, is obtained by multiplying the voltage across the load times the current through it. Obviously, if the voltage and current are out of phase, as indicated

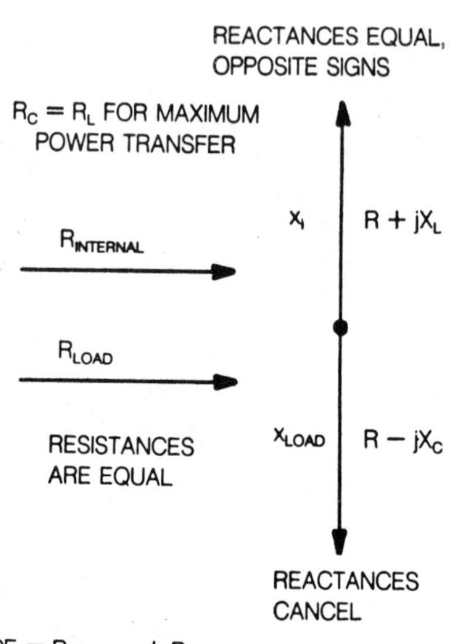

Fig. 5-16. In this illustration you can see that the reactances cancel because they are 180 degrees out of phase. The internal resistance and load resistance are in the same direction, so they add to produce the total resistance of the circuit.

by the angle θ, this will not give you the power that is actually being dissipated. To get that value, called the true power, you need to multiply $V \times I$ by the cos θ.

The power that would be dissipated by the reactance of the circuit if it were resistance is measured as VARS, (Reactive Volt Amperes). To get this value, you simply multiply $V \times I \times$ SIN θ.

Returning to true power, it is important because it represents the actual power dissipated in an ac circuit in the form of heat. (No power is dissipated by reactances.)

The value of cos θ is called the *power factor*. The power factor can be thought of as a measure of how close the true power is to the apparent power. Or, the maximum possible power will only occur if the phase angle is zero degrees, meaning that the power factor is 1.0.

The COSINE of 0 degrees is 1.0. This is the ideal power factor. So, when you know the power factor, you can tell from that value how close you are to operating with an phase angle.

It is very important in industrial electronics to pay attention to the voltage-current phase angle. The value in VARS actually represents the amount of power that is borrowed from the system on one half-cycle and returned on the next half-cycle. This has a terrible effect on the internal operation of the generator.

When the power in the generator consists of the power being generated plus the power that is being returned, the total power can be sufficient to destroy the generator. At the very least, the internal heat can produce serious damage. In industrial systems the power factor is closely monitored.

Resistive Matches. Figure 5-17 shows a generator connected to a load resistance through a matching T network. It is assumed that the phase angle is so small it can be disregarded.

The purpose of any impedance matching network is to permit the generator to "see" its internal impedance value and at the same time permit the load resistance to "see" its resistance value. So, looking in from the two ends, the impedances appear to be matched. That is the desired condition.

Instead of the T matching network of Fig. 5-17, the *delta*, or *pi* matching network in the same illustration can be used. There is a basic mathematical procedure in resistive networks called the *delta-wye transform*. It says that by using the proper calculated values, the *T* network can be replaced by the *pi* network, and, the *pi* network can be replaced by the *T* network in any circuit. The impedances looking inward from the terminals would be identical after a transformation has been made.

You will see *T* and *pi* impedance matching networks used in low power systems. Attenuators are resistive matching networks when the amount of signal loss can be varied without affecting the impedance match. The amount of signal attenuation in matching pads or attenuators is called the insertion loss.

Transformer Impedance Matching. For high power systems and systems that are more efficient than resistive networks, a transformer is used. You will remember that the impedance ratio of transformers is given by the following equation:

$$\frac{N_s}{N_p} = \sqrt{\frac{Z_s}{Z_p}}$$

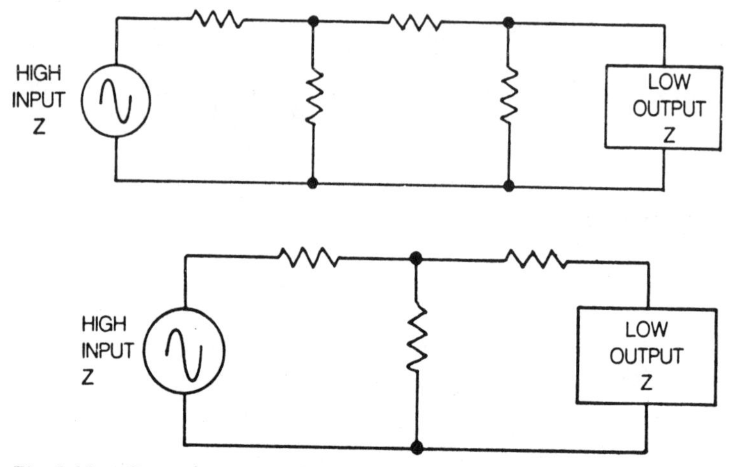

Fig. 5-17. A T-matching network.

By using the proper transformer, the load impedance will "see" its impedance value and the generator impedance will "see" its internal impedance. Therefore, the system will be matched and maximum power transfer can be obtained.

In the discussion so far, we have talked about an internal impedance of an electromechanical generator. In electronic systems, the generator can be an oscillator amplifier, or any number of circuits other than an electromechanical generator.

Amplifiers as Impedance Matching Devices. In some cases an amplifier may be used to match impedances. This may give an additional advantage of having an amplifier interface. Two circuits are of interest. One is called the follower, illustrated in Fig. 5-18. In that circuit, the input signal is to the base of the transistor and the output signal is from the emitter. This circuit will match a high impedance to a low impedance.

Instead of a transistor, a FET or a vacuum tube can be used. In these cases they are called source followers or cathode followers respectively. The one shown in Fig. 5-16 is called an emitter follower.

To match a low impedance to a high impedance, the circuit in Fig. 5-18 can be used, but it is not often used for impedance matching. Remember that the same circuit will work for tubes and field effect transistors. With a bipolar transistor it is called a

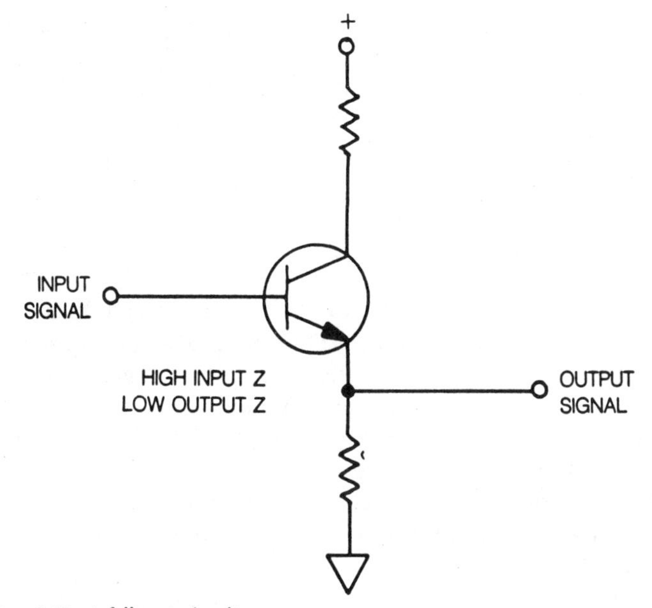

Fig. 5-18. A follower circuit.

common-base amplifier. In the case of tubes and FETs it would be called common grid or grounded grid amplifier, and common or grounded gate in a FET circuit.

You will often see this circuit used in a high frequency operation because the control electrode (base, gate, or grid) acts as a shield between the input signal and the output signal. So, the signal is not short circuited by the internal capacitance of the amplifying device.

Keep in mind that impedance matching is accomplished for the simple reason that maximum power can be transferred through matching impedances. If maximum power is not desired, then the tradeoff between efficacy and maximum power transfer can be chosen a different way.

SUMMARY

As an industrial electronics technician you will be involved with power in various forms. An important part of industrial electronics is conversion of power from one form to another.

This can be done with small units or very large units, but the principle is the same in either case.

The four methods of converting power are by:

- Converters
- Inverters
- Rectifiers
- Transformers

Another important function of industrial electronics is to make the maximum possible use of power. Maximum transfer of power is not always desirable, because efficiency is only 50 percent under maximum power conditions. However, in some cases it is nonetheless necessary to go all out and get as much power from a power source as possible.

In dc systems it is only necessary to match the load resistance to the internal resistance of the supply to get maximum output power. In ac systems a conjugate match is necessary. Actually, a conjugate match puts the load impedance in resonance with the internal impedance of the generator.

Impedance triangles and power triangles are convenient ways of representing conditions in an ac circuit. Be sure you understand what those triangles tell you about circuit conditions.

Transformers can be used for impedance matching. If there is no phase angle problem, the impedance can be matched by *T* or *Pi* resistive networks. *T* networks are also called wye or Y networks. Delta networks are also called *Pi* networks.

SELF TEST

1. Which of the following converts dc to ac?
   (A) rectifier.
   (B) inverter.
   (C) converter.
   (D) transformer.

2. In which of the following is an oscillator used?
   (A) inverter.
   (B) converter.
   (C) Both choices are correct.
   (D) Neither Choice is correct.

3. In a transformer, the secondary impedance ratio is equal to:
   (A) the primary turns ratio.
   (B) the square of the primary turns ratio.

4. In a typical operation of a self-saturating transformer with a sine wave input to the primary, you would expect the waveform across the secondary to be:
   (A) a square wave.
   (B) a sine wave.

5. Which of the following components is necessary for making a ferro-resonant transformer operational?
   (A) resistor.
   (B) capacitor.
   (C) constant-current diode.
   (D) None of these choices is correct.

6. What is the conjugate of $15 - j4$?

7. The amount of power that appears to be dissipated by an inductor in an LR circuit is measured in:
   (A) watts.
   (B) volt-amperes.
   (C) VARS.
   (D) None of these choices is correct.

8. Apparent power is measured in:
   (A) watts.
   (B) volt-amperes.
   (C) VARS.
   (D) None of these choices is correct.

9. Which of the following is used to determine the power factor of an ac circuit?
   (A) SIN θ.
   (B) COSN θ.
   (D) TAN θ.

10. T network can be replaced with a *Pi* network by using
    (A) delta-wye transform.
    (B) wye-delta transform.

ANSWERS TO SELF TEST

1. (B)
2. (C)
3. (B)
4. (A)
5. (B)
6. 15 + j4
7. (C)
8. (B)
9. (B)
10. (B)

6

Power Supplies

ELECTRONIC COMPONENTS THAT are used for amplifying, logic circuitry, microprocessors, computers, and other control systems that operate electronically require a dc voltage for their operation. A battery can be used to supply the dc, but the usual approach is to use a power supply.

A power supply operates from the ac power line. The term power supply is a misnomer. It does not really supply power. The power source—usually the ac power line—does that. Instead, the power supply modifies the input ac power to provide dc operating voltages for a system. Rectifiers are a very important part of the power supply. A portion of this chapter will also review rectifier operation.

In a *brute force* power supply, there is no regulation of the dc output voltage. So, the output voltage varies with the load resistance. In the ideal supply there would be no variation. The percent regulation is a measure of how closely the brute force supply approaches the ideal condition.

Regulated supplies are very important to some types of operation. This type of supply can be divided into two groups: analog and switching.

Analog regulators provide continuous control of the dc output voltage. Switching regulators control the output voltage by breaking up the dc voltage into pulses. The RMS value of the pulse is controlled to regulate the output power of the supply.

Filtering circuits are used to obtain a very smooth output voltage. In the early days of electronics, filters consisted of simple combinations of capacitors and coils or resistors. Today filters may consist of very complicated combinations of R, L, and C. These combinations are computer designed and they are very difficult to analyze by chasing electrons through the circuit.

Filters may also be made from simple electronic voltage regulators. In that case they are called electronic filters. The most popular circuit for this type of filter is discussed in this chapter.

Some types of electronic equipment can be severely damaged by an overvoltage or overcurrent. If something goes wrong with the electronic regulator, an overvoltage or overcurrent condition can exist. To prevent that, special circuits are sometimes employed to guard against excessive output. Examples are foldback current limiters and crowbar circuits.

Voltage dividers may be used at the output of a power supply. Remember, however, that voltage dividers can eliminate the advantage of regulated supplies. For example, if the output voltage is taken from a divided voltage, the effect is to put a resistance in series with the output. That resistance behaves like the internal resistance of a supply and defeats the purpose of regulation.

CHAPTER OBJECTIVES

Here are some of the questions you will find answered in this chapter:

- What is percent regulation and how is it calculated?
- How does an analog regulator work?
- How does a switching regulator work?
- What components are used in a crowbar circuit and how does that circuit work?
- What are the characteristics of a three-legged regulator?

DIODE CONFIGURATIONS

Before discussing some basic rectifier configurations, we need to review some diode connections. Figure 6-1 shows three diodes: one that is forward biased, one that is reverse biased, and one with no bias (for comparison). These are, of course, models to illustrate the effect of bias on the size of the depletion region.

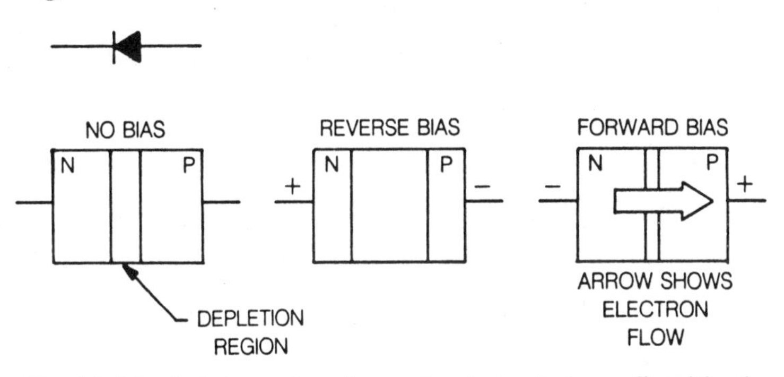

Fig. 6-1. The depletion region of a semiconductor diode is affected by the polarity of voltage across it.

Note that in the forward bias diode the depletion region is reduced to a very small cross sectional area. When the diode is reverse biased, the size of that depletion region is greatly increased.

Even though diodes may have exactly the same number or identification, the area of depletion zone in both the forward and reverse condition can vary widely from one diode to another. One way of visualizing the diode is that it is basically a capacitor. This is especially true when they are reverse biased.

Figure 6-2 shows how a reverse voltage across a diode affects its capacitance. The P and N regions serve as conductors and the depletion region serves as the insulator. As shown in the illustration, increasing the reverse bias moves the plates apart. That, in turn, reduces the capacitance.

Figure 6-2 shows that a semiconductor reverse biased diode can be used as a capacitor. When they are specifically designed for this purpose they are called *varactor diodes*. They are also sometimes called by the trade name varicap.

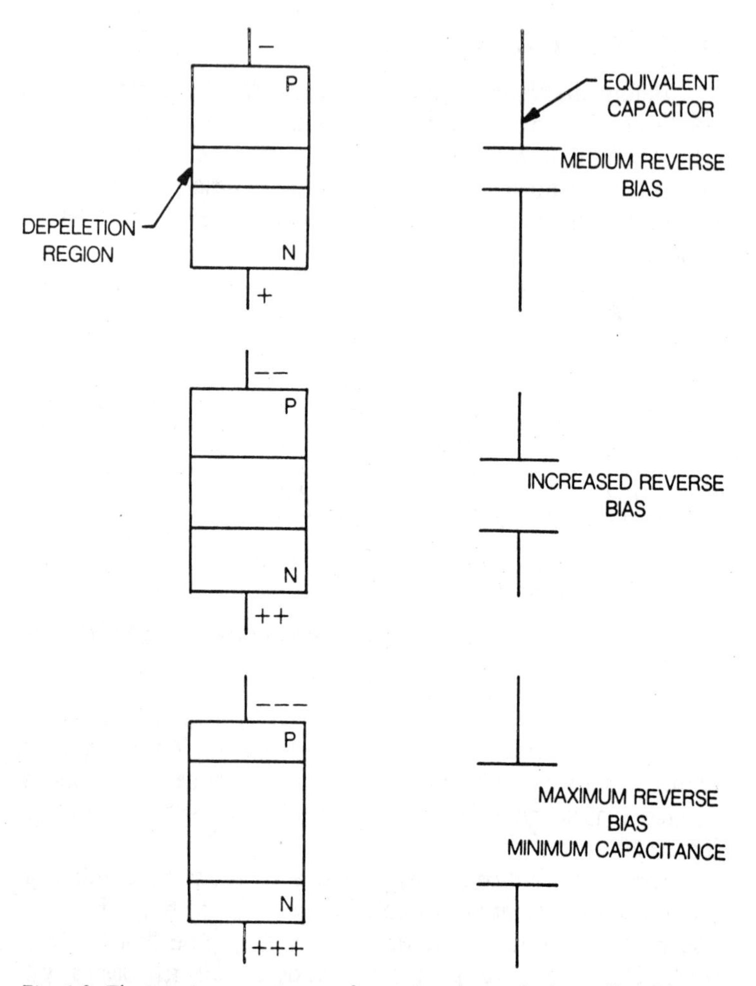

Fig. 6-2. The junction capacitance of a semiconductor diode is affected by the amount of reverse voltage across it. Note that higher reverse voltages result in lower capacitance values.

It is important to understand the relationship between capacitance and voltage when rectifier diodes are connected in special arrangements. Figure 6-3 shows one relationship by representing reverse biased diodes as being capacitors.

The higher voltage is across the lower capacitance value. If there were three or more different sizes of capacitance in series,

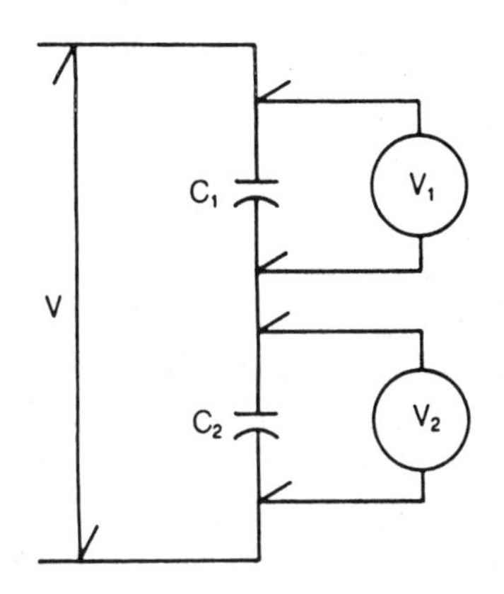

The capacitor charge of any capacitor is given by the equation

$$Q = CV$$

For each capacitor

$$Q_1 = C_1{}^{V}{}_1$$
$$Q2 = C_2{}^{V}{}_2$$

But, since they are in series, both capacitors must have the same amount of charge. ($Q_1 = Q_2$) So,

$$C_1V_1 = C_2V_2$$

The voltages add to equal the applied voltage V.

$$V = V_1 + V_2$$
$$\text{or } V_2 = V - V_1$$

By substitution

$$C_1V_1 = C_2(V - V_1)$$

Solve for V_1

$$V_1 = \frac{C_2V}{C_1 + C_2}$$

By a similar solution:

$$V_2 = \frac{C_1V}{C_1 + C_2}$$

Fig. 6-3. When semiconductor diodes are connected in series, their junction capacitance values are also in series. This will affect the peak inverse voltage across the diode as illustrated in this example.

Example:

$C_1 = 0.001$
$C_2 = 0.0015$
$V = 25V$

$$V_1 = \frac{0.001 \times 25}{0.001 + 0.0015} = 10V$$

$$V_2 = \frac{0.0015 \times 25}{0.001 + 0.0015} = 15V$$

R_L	P_L
0	0
2	5.55
4	6.25
6	6.0
8	5.55
10	5.1

R_L	% efficiency
0	0
2	33.3
4	50
6	60
8	66.7
10	71.4

Fig. 6-3. cont

it would also be true that the highest voltage is across the smallest capacitance.

Series Diodes. Figure 6-4 shows how three diodes are connected in series to increase the peak inverse voltage (PIV) rating. The PIV rating is the maximum voltage that can be connected across a diode in the reverse direction without breakdown.

For the three diodes in series in Fig. 6-4, the peak inverse voltage rating is equal to the sum of each rating. If they are identical diodes, then the peak inverse rating is equal to three times the rating of one diode.

Disregard the capacitors and resistors across these diodes for the time being. When the voltage across the diodes is in the reverse direction, each diode has a junction capacitance. That

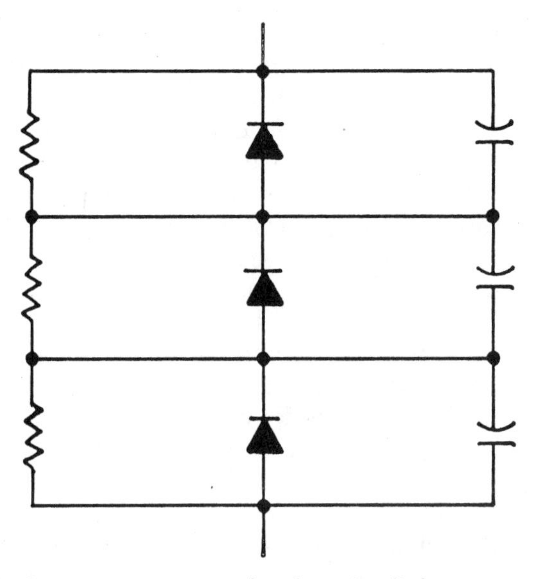

Fig. 6-4. This connection assures that the series diodes are protected from uneven reverse voltages.

presents no problem if all of the junction capacitance values are identical. In real life there will be three different junction capacitance values for the three diodes.

From Fig. 6-3, you know that the highest voltage value will be across the lowest junction capacitance value. If the reverse voltage is high, as would be indicated by putting the diodes in series to get a high PIV rating, then you can expect the diode with the lowest junction capacitance to break down before the other two. Remember, the breakdown voltage of one of the diodes can exceed its rating if it has a small junction capacitance compared to the other two.

In Fig. 6-4, there are three capacitors connected in parallel across the diodes. These capacitance values are large compared to the junction capacitance. In fact, a ten to one ratio is assumed. Furthermore, the capacitors are matched so that when a reverse voltage occurs across the combination, each will have exactly the same amount of reverse voltage. This has the effect of eliminating the influence of junction capacitance on the reverse voltage.

Rectifier diodes are not perfect devices. When they are reverse biased there will be some leakage current through each diode in Fig. 6-4. Because of this reverse leakage current, there will be a voltage across each diode resistance. If the opposition to the reverse current were the same for each diode, this will not present a problem. However, reverse resistance values can vary over a relatively wide range. This means that the reverse voltages will not be evenly distributed between D1, D2 and D3.

However, the resistors in parallel with these diodes have identical high resistance values. The high resistance values equalize the reverse voltages across the diodes. In some circuits you will see only the capacitors or only the resistors.

Parallel Diodes. Consider now the problem of connecting diodes in parallel as shown in Fig. 6-5. This is done to get a greater forward current rating. Disregard temporarily the fact that there are two series resistors in the circuit.

As you might expect, the forward resistance of the diodes varies over a range of values, even though the diodes have the same identification number. So, when the diodes are forward biased, one will have a greater forward voltage drop than the other. The diode with the lower forward voltage will prevent the other diode from getting into conduction.

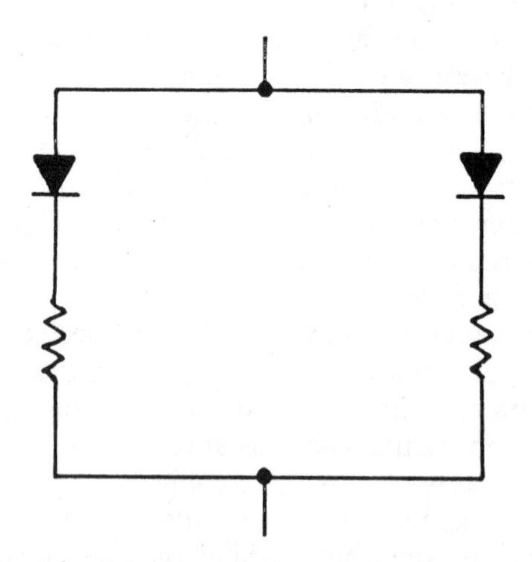

Fig. 6-5. Current hogging is prevented by including the resistors in this parallel diode connection.

Suppose, for example, one diode requires a forward voltage drop of 0.7 volts when conducting and the other one has a forward drop of 0.6 volts. If the 0.6 volt diode begins to conduct, it will not be possible to get the 0.7 volt required for the other diode. So, all the current will flow through the diode with the lower forward resistance. This is called current hogging.

Now return to the circuit with the two series resistors. The resistance values are low, but they are large values compared to the forward resistances of the diodes. So, a voltage drop in each parallel branch is higher than the voltage required for starting the opposite diode. Regardless of which one starts, the voltage across that parallel branch will be sufficient to start the other diode. To summarize, the resistances in series with the diodes in the parallel circuit are used for preventing current hogging.

Parameters. Parameters are values you choose in order to make a problem come out right. If someone tells you to draw a square having a specified number of square inches of area, the parameter you choose is the length of the side. Specifications for electronic devices are based on the parameters of operation for those devices. Of specific interest at this time is the diode. What are the parameters that determine its operation?

One parameter is the amount of forward current that the diode can conduct without becoming overheated and destroyed. Another important parameter is the amount of reverse voltage that can be tolerated across the diode before it breaks down. There is no such thing as a perfect insulator. If you put a sufficient amount of voltage across a diode in the reverse direction (positive cathode, negative anode), current will flow. The amount of voltage required is called the breakdown or peak inverse voltage. (This is also true of a vacuum tube diode.) When the diode conducts during a reverse breakdown, it is usually destroyed. There are some exceptions.

Specifications are based on the parameters of device operation. If you look at diode specifications, you will find the maximum forward current and the maximum allowable reverse voltage that can be used for that particular diode. Those are the two things you need to know if you are going to buy a diode for a replacement. Because of different specifications, you cannot use just any diode in an application. It must have the correct parameters, or specifications. It is always best to use an exact diode replacement.

A Characteristic Curve. Not all semiconductor diodes are made with silicon. Some are made with germanium. Others are made with gallium arsenide (GaAs). The characteristic curves are different for each device.

Figure 6-6 shows a characteristic curve of three types of diodes. An important feature of this characteristic curve is that no current flows, for all practical purposes, until a certain voltage is reached. For a germanium diode, that voltage is usually around 0.2 to 0.3 of a volt. For silicon diodes it is about 0.7 volts or higher. For a gallium arsenide diode the voltage must be 1.5 volts or greater. From these three conditions you can readily see that you cannot put one diode in another diode's place if its junction voltage is not correct.

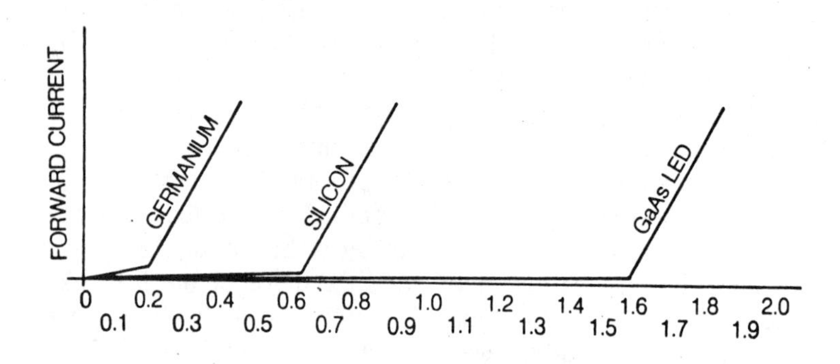

Fig. 6-6. The forward breakover voltages of three types of diodes are graphed here. Gallium arsenide (GaAs) is used to make light-emitting diodes and very-high-frequency transistors.

The fact that junction voltages exist gives us one method of testing P-N junctions. Using a voltmeter with a high impedance, we can measure the forward voltage drop across the junction. If that voltage does not exist, the diode is not conducting. That may be because the diode has been destroyed, or it may be because the external circuit is not delivering the proper voltage.

RECTIFIER CIRCUITS

Rectifiers are covered in basic courses that are prerequisite for this book. They will briefly be reviewed here for convenience.

Single Phase. The simplest half-wave rectifier circuit is shown in Fig. 6-7. This circuit is transformer driven. If the ac voltage on the power line is the proper value, the transformer can be eliminated. However, this circuit has the advantage that the secondary circuit is "floating." In other words it can be grounded at any point. This is a safety feature, because if a person touches the circuit at the same time he touches ground, the circuit will float to the ground potential and the person will not be seriously injured.

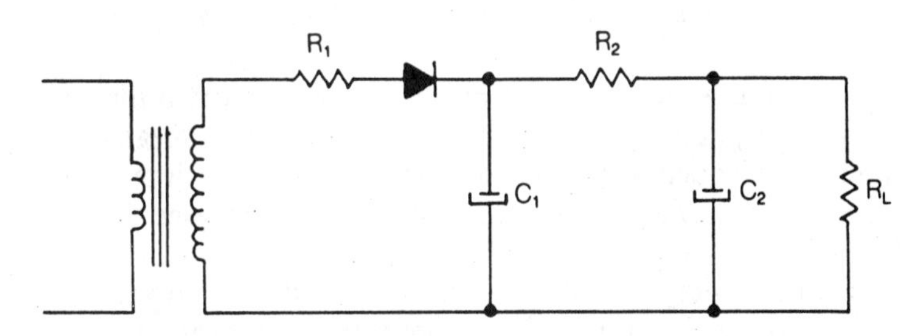

Fig. 6-7. This simple half-wave rectifier circuit has a surge limiting resistor (R1) to protect the diode.

Without Resistor R1, the charge of the electrolytic capacitors on the first half-cycle can be sufficiently high to destroy the diode. In practice R1 will be a value less than 10 ohms.

The RC filter circuit uses a resistor instead of a choke. The advantage of this is that it is much cheaper and requires less space. The disadvantage is that the voltage drop across the filter resistor reduces the available output voltage.

The ripple frequency of a half-wave power supply is equal to the line frequency. So, a 60 hertz ac power line will result in a ripple frequency of 60 hertz. Despite the disadvantage of greater ripple, half-wave rectifiers are used extensively in high-voltage circuits. They are also popular in low-budget circuits where the transformer is not needed.

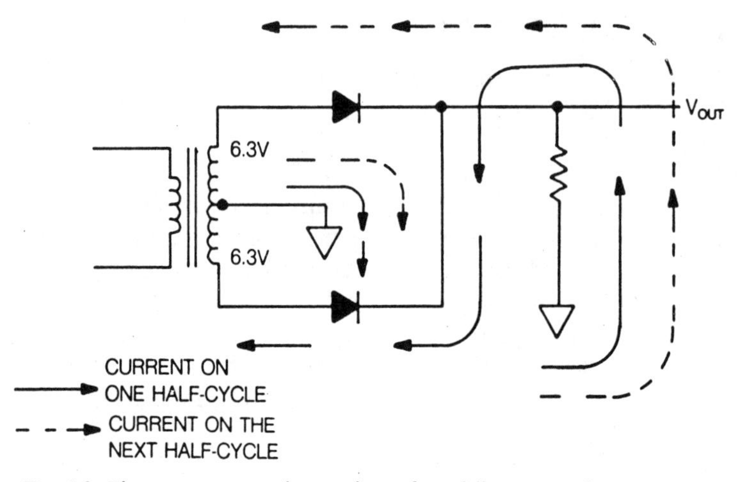

Fig. 6-8. Electron current paths are shown for a full-wave rectifier circuit.

The full-wave rectifier in Fig. 6-8 provides current through the load (R_L) on both half cycles of ac input. This means that the ripple frequency will be 120 hertz for a 60 hertz input. The ripple frequency of a full-wave rectifier is always twice the line frequency.

An advantage of full-wave rectifiers is that the higher frequency ripple is easy to filter compared to the output of a half-wave rectifier. A disadvantage is that a tapped secondary winding is required. That means that you can use only one half of the total secondary voltage to produce an output.

The bridge rectifier circuit of Fig. 6-9 is also a full-wave rectifier. Compare it with the full-wave rectifier in Fig. 6-8. Note that the total output voltage is utilized to produce a voltage across R_L. This is also true of the half-wave rectifier, but the lower ripple frequency is a disadvantage in that case. In the earlier days of electronics, the use of four diodes was considered to be a disadvantage of the bridge rectifier. In those days the diodes were vacuum tubes and were the most unreliable part of the circuit. Today, the highly reliable semiconductor diodes make the bridge an ideal circuit for many full-wave applications.

Doublers. Figure 6-10 shows a half-wave voltage doubler circuit. You can follow the two half-cycles with the arrows. On the first half-cycle, capacitor C1 is charged to the peak voltage

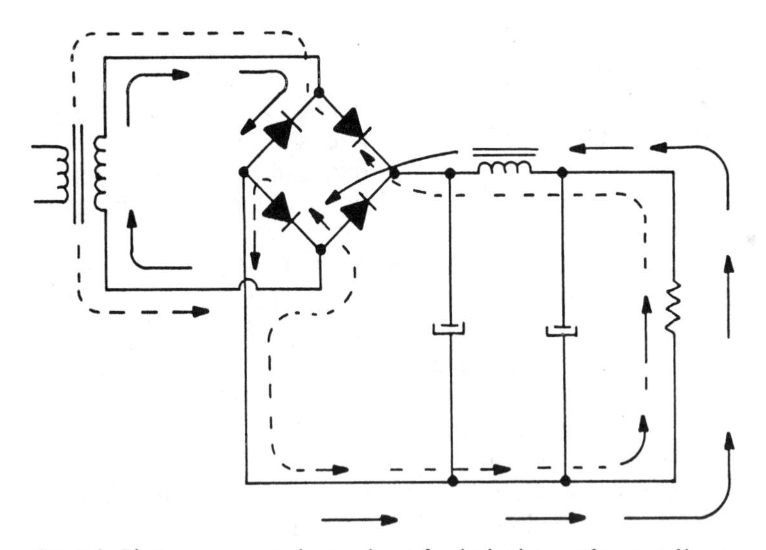

Fig. 6-9. Electron current paths are shown for the bridge rectifier. A pi filter is used to smooth the output of this full-wave rectifier.

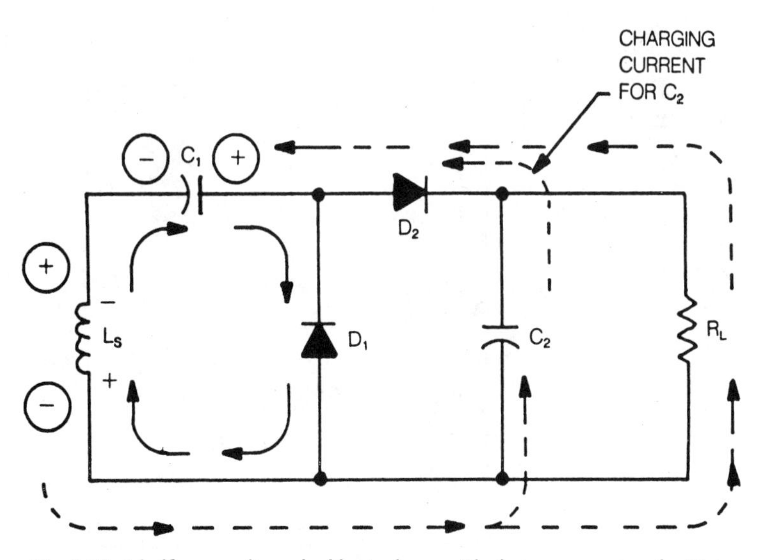

Fig. 6-10. A half-wave voltage doubler is shown with electron current paths. Note that current flows through the load resistance for one-half cycle only.

through diode D1. On the next half-cycle the voltage across the secondary of the transformer is in series with the voltage that was established across C1 on the previous half-cycle. Observe that the two voltages, which are circles in the diagram, add to produce the output voltage. This results in a total voltage across the rectifier circuit that is twice the peak value of the voltage for one cycle.

Capacitor C2 will charge to the peak voltage. It is used for storing the output voltage and for filtering the output. Note the charge current for C2 is through D2 on the first half-cycle. This can be a very undesirably high current, but it only occurs until C2 is fully charged. Remember that current flows through R_{LOAD} only for one half-cycle.

Figure 6-11 shows a full-wave voltage doubler. The first half-cycle causes C1 to be charged to the peak value through D1. On the second half-cycle, C2 is charged to the peak value through D2. Note that the voltages across the two capacitors are in series, so they add to produce an output voltage across the load resistor that is twice the secondary voltage.

The question is sometimes asked: Since the full-wave doubler takes exactly the same number of components as the

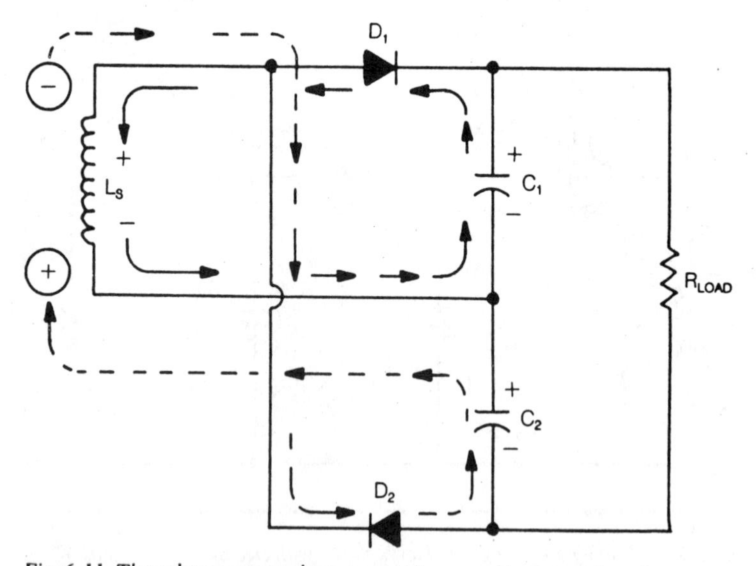

Fig. 6-11. The voltages across the two capacitors combine to produce the output voltage of this full-wave doubler.

half-wave doubler, but the full-wave doubler has an output that is easier to filter, are there any reasons for using the half-wave doubler?

The answer can be seen by comparing the two circuits. Notice that in the full-wave doubler, the bottom of the secondary winding is at a voltage halfway between the voltage across C2 and the voltage across C1. That means that the bottom half of the secondary is above the secondary circuit voltage by an amount equal to the peak voltage across the secondary winding.

Half-wave doublers have a common line that can be connected directly to the ac power line. If the voltage is correct, the half-wave doubler can be connected directly to the power line without a transformer. That is not possible with the full-wave doubler.

POWER SUPPLY REGULATION

The power supplies discussed so far have been examples of brute force design. This means that the output voltage and current is the result of the choice of transformer and components, but the output is not controlled electronically. Figure 6-9 shows a good example.

One of the features of interest in brute force design is the percent regulation. This is a measure of how well the output voltage stays at a constant value when the load current changes.

When the load current changes, the current through the internal resistance of the supply changes. That in turn lowers the output voltage. This can be seen in the simple Thevenin's equivalent circuit of Fig. 6-12.

Note that the load resistance, which is shown as a variable resistor, can change the amount of current through the supply. As the current increases, the voltage across the output terminals and therefore across the load resistance, decreases. In an ideal power supply the output would be a constant voltage.

The equation for percent regulation is given here for reference:

$$\text{Percent Regulation} = \frac{V_{NL} - V_{FL}}{V_{FL}} \times 100$$

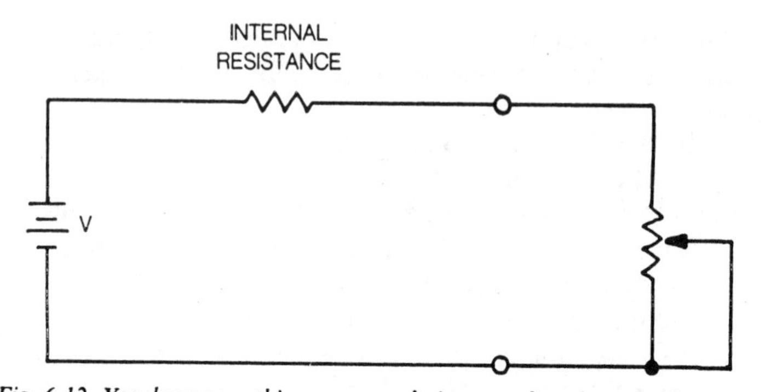

Fig. 6-12. You have seen this power supply in an earlier chapter. Changes in load resistance will affect the output voltage. When that change in output voltage cannot be tolerated, a voltage regulator is required.

where: V_{FL} is the full-load voltage
V_{NL} is the no-load voltage.

The best possible supply would have a 0 percent rating.

Voltage Regulation. There are many circuits in electronics that require a stiffly regulated power supply. Examples are microprocessors and TTL logic circuits. They require a precise five volt supply. To provide it, voltage regulators are used.

The study of voltage regulators is especially important for industrial electronics. One reason is that every industrial electronics circuit requires a power supply for its operation. Regulated dc voltages are often needed for operating these electronic devices.

Another reason for studying regulators is that they provide insight into closed-loop circuit operation. Closed-loop circuits are the foundation of many industrial electronic control systems.

Series and Shunt Regulators. All regulators can be divided into the two classes shown in Fig. 6-13. In the series regulator, all the supply current flows through the load resistance and through the regulating device. In the shunt regulator, part of the supply current flows through the load and part flows through the regulating device.

A zener diode is shown as a shunt regulator in Fig. 6-14. This is not a very important circuit for regulation as far as

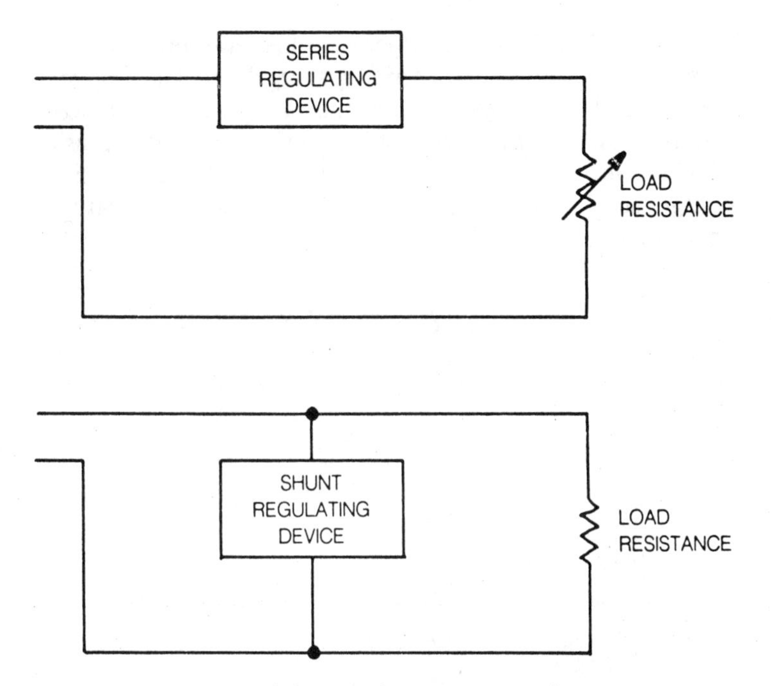

Fig. 6-13. Here, series and shunt regulators are compared.

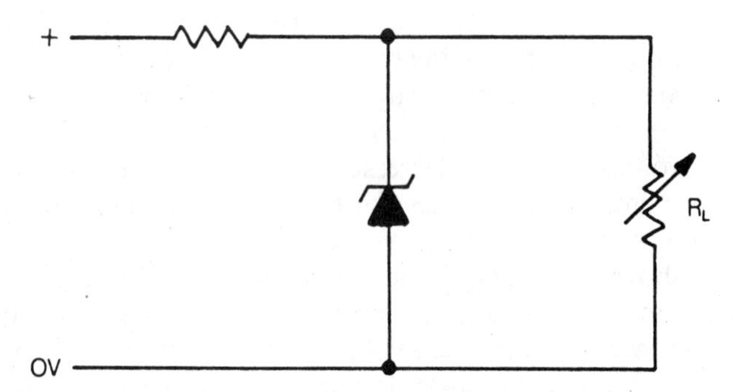

Fig. 6-14. This simple zener diode shunt regulator is often used to make a reference voltage for more complicated voltage regulators.

power supply output is concerned. But in electronically regulated power supplies you will often find a zener regulator circuit. It is used for providing a reference voltage that is used for establishing a regulated output.

Figure 6-15 shows the principle of the series regulator. The transistor is usually called a series-pass regulator. The base voltage is fixed by the adjustment of the variable resistor. Since a battery is used for the source of base voltage, it will be considered to have a constant bias. With the base fixed and the emitter voltage set by the drop across R_L, regulation is easy to understand.

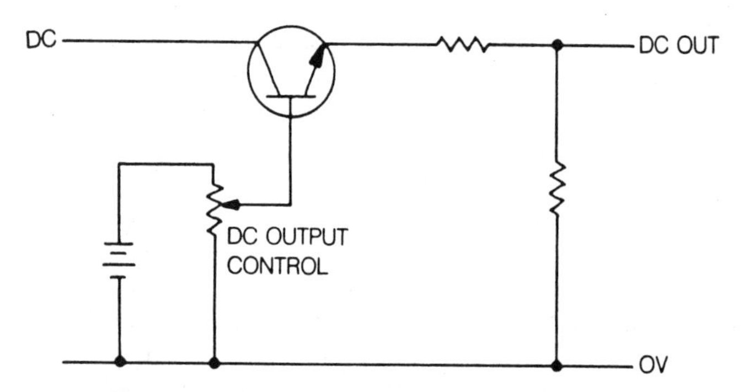

Fig. 6-15. The output voltage can be controlled by adjusting the bias of the series-pass transistor.

If the output voltage starts to rise—that is, the emitter becomes more positive—the emitter-to-base voltage decreases. That lowers conduction through the series-pass regulator and lowers the output voltage back to the desired value. If the output voltage starts to decrease, there is an emitter-to-base voltage increase. This causes the transistor to conduct harder and raise the output voltage back to the desired value.

As shown in Fig. 6-16, the series-pass and shunt regulators can be combined into a simple regulator circuit. The battery has been eliminated, but the base voltage is fixed by the zener diode.

The circuit in Fig. 6-16 is often used as a power supply filter. In this case the load resistance is the output of the complete power supply system. Capacitor C1 prevents the series-pass regulator from operating so quickly that it overshoots the desired value. Saying it another way, the system will not respond to momentary transient voltages. The series and shunt regulators are examples of unregulated power supplies. Although the output voltage indirectly controls the operation of

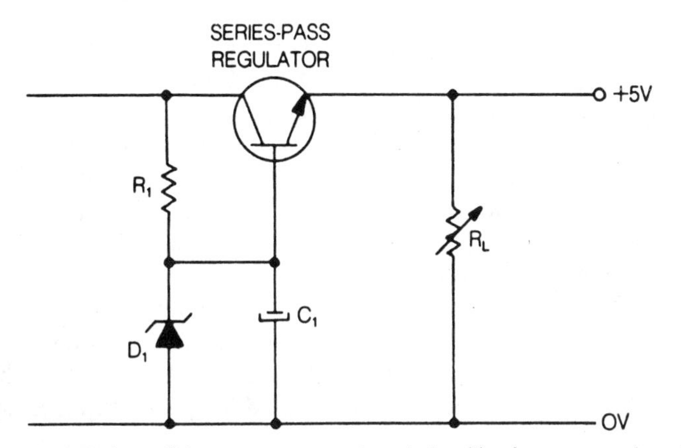

Fig. 6-16. The bias of the series-pass transistor is fixed by the zener regulator in this open-loop (no feedback) circuit.

the circuit, there is no amplified feedback system. So, they are not called closed-loop regulators.

Series regulators are by far the most popular types in use. They can be divided into two basic classifications. One is the analog regulator. This has continuous sensing and feedback of the output voltage, or current, in order to maintain the output at a constant value. In the switching regulator, the output of the supply is continually sensed. However, it does not provide full-time regulation or control of the series control element.

The switching supply has the higher efficiency. Furthermore, the high switching speed provides an output that is much easier to filter. So the filter components can be smaller and more efficient. Each of these types of regulators will be discussed with reference to a sample circuit.

SCR AND CLOSED-LOOP REGULATORS

You will remember that an SCR is basically a rectifier. But is different from other rectifiers in that it can be switched on almost instantaneously by using a gate current. Until the gate current flows, no cathode-to-anode current flows. The SCR makes it possible to regulate the output voltage from any of the single-phase rectifiers that we have discussed.

A bridge rectifier with SCR control is shown in Fig. 6-18.

Remember that the same theory applies to any rectifier circuit. So by using an SCR and the proper firing voltage, it is possible to regulate the output of all rectifier circuits.

In any closed-loop system, part of the output is fed back and used to control the input. Closed-loop voltage regulators for power supplies are good examples. They sample the output voltage and feed back a signal (usually amplified) that is used to control the output voltage.

Other examples of closed-loop circuits are: motor speed controls using phase-locked loops, automatic frequency controls for oscillators, and position controls used for robots. The two types of closed-loop regulators used for power supplies are analog and switching.

The SCR Bridge. Only two SCRs are used in the SCR regulator. There is one SCR for each half-cycle of output. In some cases you will see that all four diodes in the bridge are SCRs. However, in the arrangement of Fig. 6-17, one SCR for each half-cycle is sufficient because the other diode cannot conduct if the SCR is off.

As shown in the illustration, each SCR conducts after its half-wave input has started. This removes part of the output power and lowers the RMS value of the output voltage. The amount of each half-wave that is cut off depends upon the time at which the gate of the SCR is triggered. The firing time can be

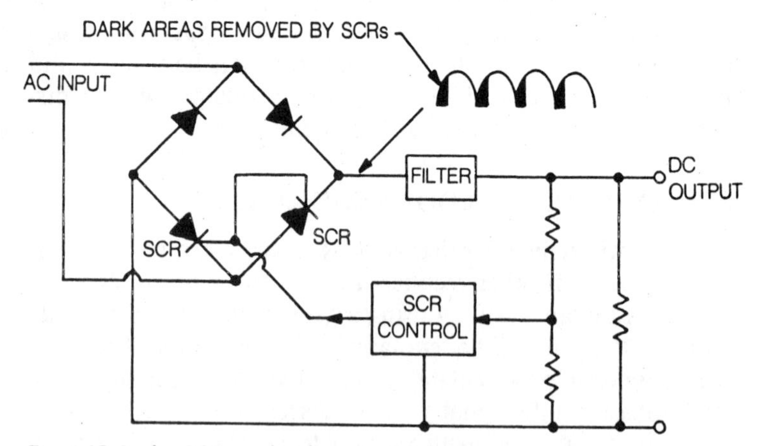

Fig. 6-17. In this SCR regulated supply, the thyristors control the output power.

controlled by circuits that will be discussed later in this chapter, or it can be set with a manual control.

Analog Regulators. Figure 6-18 shows a block diagram of an analog regulator circuit. The power source is the commercial ac power line. Following the input there is a rectifier and filter. The output of the rectifier and filter is an unregulated dc voltage. The rectifier and filter in this circuit is a brute force power supply, which may be either half-wave or full-wave, but if it is important enough to go to the expense of a closed-loop regulator, it can be presumed that the rectifier will be a full-wave type. Bridge rectifiers are very popular for this type of system.

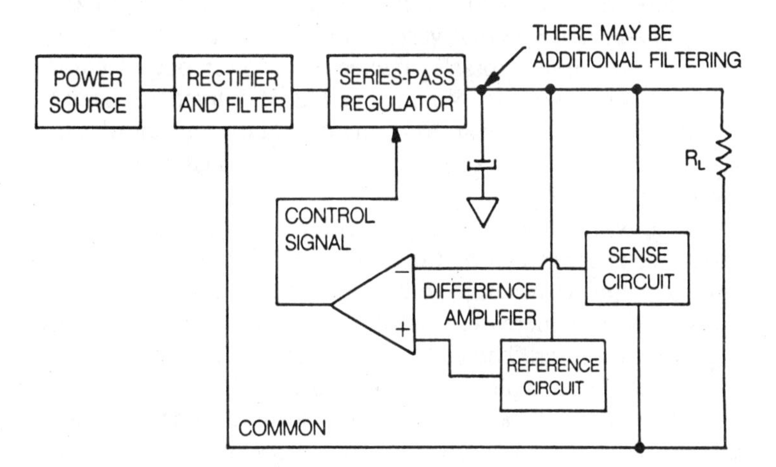

Fig. 6-18. A simple analog voltage regulator circuit.

The series-pass regulator is a power amplifier. It controls the amount of current that flows through the load resistance (R_L). There are two circuits across the output line of the power supply. One provides the dc voltage delivered to the difference amplifier. The other is called the *sense circuit.*

In its simplest form, a sense circuit is a voltage divider that produces an appropriate dc voltage for the inverting input of the difference amplifier. The difference amplifier is shown to be an op amp in this case, but any voltage amplifier can be used. The amplifier is also called a sense amplifier or differential amplifier. It delivers a dc output voltage that is directly related to the difference in the reference voltage and sense voltage.

In the zero-correction condition, the control signal delivered to the series-pass regulator sets the voltage across R_L at the desired value. If that voltage tends to increase, there will be a more positive voltage out from the sense circuit. When delivered to the inverting input of the amplifier, a negative-going output signal goes to the series-pass regulator. Conduction of the series-pass regulator is decreased, lowering the output voltage until it is back to the required value.

Note that even though there is an increase in the output voltage to the supply, the reference voltage is not changed. Therefore, the inputs to the differential amplifier consist of a fixed voltage and one that varies with the output of the supply.

If the output across R_L decreases, a lower output comes from the sense voltage. When inverted, it delivers a positive-going signal to the series-pass regulator and the current through R_L is increased until the output voltage is again set to the required value.

One way of looking at the series-pass regulator is that its resistance is controlled by the input control signal. If the input control signal is negative-going, as it would be if the output voltage tried to increase, then the resistance of the series-pass regulator increases. The greater drop across the series regulator lowers the output voltage to the required value.

Conversely, if the input signal is positive-going, the resistance of the series-pass regulator decreases, there is less drop across that regulator, and the output voltage across R_L is increased back to the desired value. It takes time to read about the action of this regulator but its operation can be very quick. In fact, its action is measured in microseconds.

Figure 6-19 shows an inexpensive closed-loop analog regulator. Here the series-pass regulator is shown as a single power transistor; however, Darlington transistors are often used in this position.

The sense circuit consists of R2 and R3 while Q2 is the sense amplifier. Its emitter voltage is fixed by the zener diode. Any change in the output voltage is felt on the base of Q2, a voltage amplifier. Resistor R1 is the load resistance for that amplifier. Since it is a common-emitter amplifier (input signal

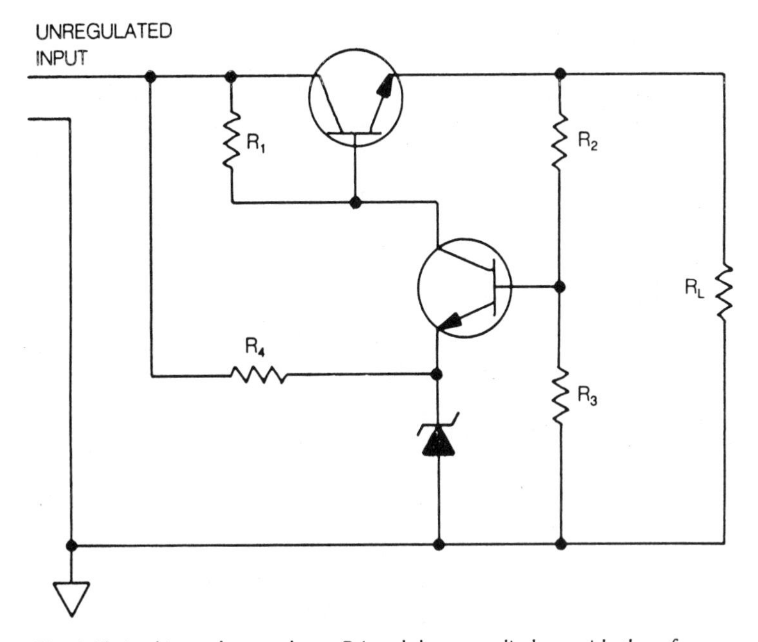

Fig. 6-19. In this analog regulator, R4 and the zener diode provide the reference voltage. Resistors R2 and R3 provide the sense voltage. Q1 is the series-pass regulator and Q2 is the sense amplifier.

at the base and output at the collector) it follows that there will be a 180 degree phase shift in the output signal. So, a positive-going input to the base of Q2 results in a negative-going signal at the base of Q1. Conversely, a negative going signal at the base of Q2 will cause the base of Q1 to go in a positive direction.

This type of inexpensive regulator is sufficiently fast to filter out any ripple that arrives from the unregulated supply. Even so, there may be additional filtering at the emitter of Q1.

Switching Regulators. To begin, we will start with the waveform shown in Fig. 6-20. These pulses represent a method of varying the RMS value and the average value of a signal. The pulse waveform at the top is a perfect square wave. This is the desired signal. In other words, it represents the desired average output value. In a true square wave, the average value is half the peak value.

When the width of the pulses is decreased, as shown in the next waveform, the output power is reduced. Both the average

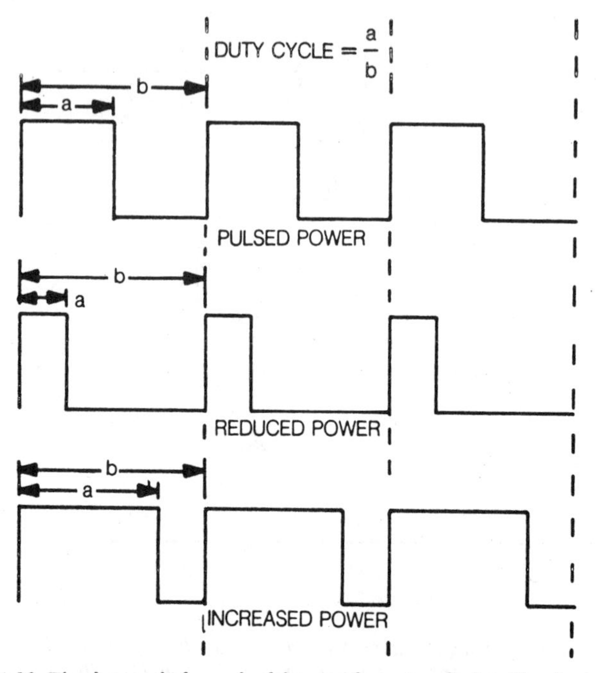

Fig. 6-20. The duty cycle for each of the waveforms is calculated by the equation. Observe the difference between increased power and reduced power.

and RMS values are decreased. If you would filter the output of this waveform and the output of the standard waveform at the top, you would find that the resulting dc would be a lower value in this case.

You can increase the output dc by increasing the width of the pulses as shown in the lowest waveform. Here, the output pulse is wider than the standard pulse power. This results in a higher average value, a higher RMS value, and a higher dc output after filtering.

The waveforms in Fig. 6-20 can be used as the input of a series-pass regulator to control the dc output of a regulated power supply. Figure 6-21 shows a series-pass regulator with these waveforms delivered to the base. As mentioned before, if these waveforms are varied as in Fig. 6-20, the output power to R_L can be controlled.

Figure 6-22 shows a block diagram of a typical switching regulator. As with the analog supply, there is a power source

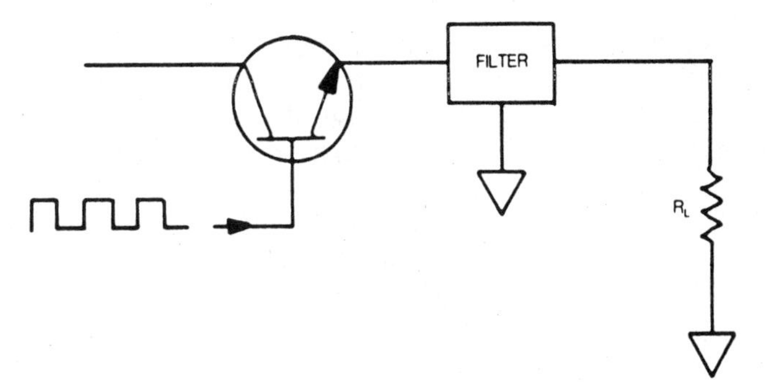

Fig. 6-21. Pulses can be used with a series-pass regulator to control the amount of power delivered to R_L

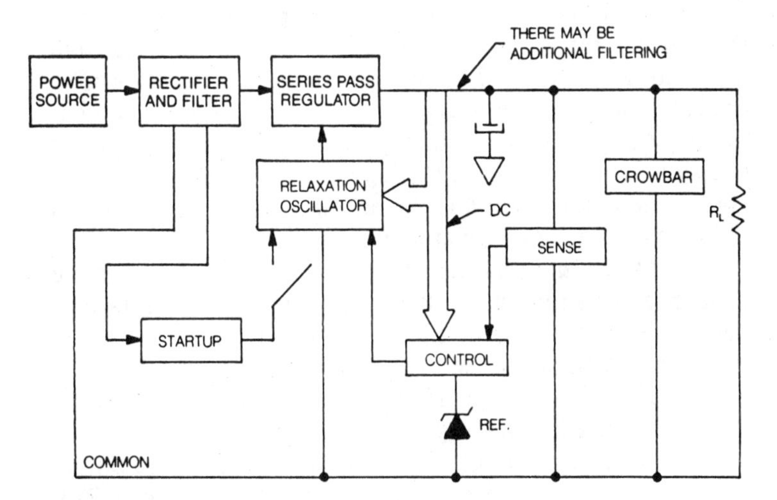

Fig. 6-22. The essential parts of a switching regulator.

(the commercial ac power line) and a rectifier with filter. That output is delivered to the series-pass regulator. It is the unregulated output of the rectifier and filter that is to be controlled.

A relaxation oscillator delivers square waveform to the base of the series-pass regulator. The pulse width of the relaxation oscillator is controlled by a pulse width control circuit. That control circuit monitors the sense input and compares it to the reference. The output of the control circuit sets the pulse width of the relaxation oscillator. If the output voltage tends to increase, the closed-loop circuit decreases the pulse width of the relaxation oscillator and lowers the output to the desired value.

If the output tends to decrease, the feedback circuit increases the pulse width and increases the output across R_L to the desired value.

The term duty cycle is often used to refer to the amount of energy in a pulse. It is determined as follows (See Fig. 6-20):

DUTY CYCLE = ON TIME/TOTAL PULSE TIME

Note that the relaxation oscillator operates from the dc output of the supply. When the system is first turned on, there is no dc output, so there is no way to control it. To eliminate this problem, a start-up circuit is used. It simply provides an input that permits the series-pass regulator to conduct and start the closed-loop operation. The start-up circuit is usually a simple oscillator, whereas the relaxation oscillator may be a multivibrator or any oscillator with a pulse output. Once the power supply is in operation, the startup circuit is automatically disabled. The sense circuit and the reference circuit are similar to those used in analog regulator systems.

The switching regulator is more complicated and more expensive than the analog type, but it has the advantage of being much more efficient. Electronic circuits are like people—if you give them a chance to rest, they work more efficiently. Since the circuit works on pulses, rather than from a continuous dc voltage, its efficiency is greater than for analog systems.

The Crowbar Circuit. You will notice that at the output of the power supply in Fig. 6-22, there is a crowbar circuit. This is an overvoltage protection circuit that prevents excessive voltage across R_L. As in all other illustrations for power supplies, R_L represents the load resistance. That is usually some kind of dc circuit or system that requires a steady dc for its operation. It is usually some kind of electronic circuitry. If the output voltage starts to rise, the electronic circuitry can be damaged. To prevent that, the crowbar circuit short circuits the output and prevents that damage.

Figure 6-23 shows an example of a crowbar circuit. It is comprised of an SCR that receives a signal from a zener diode in series with a resistor. The zener-resistor combination is the

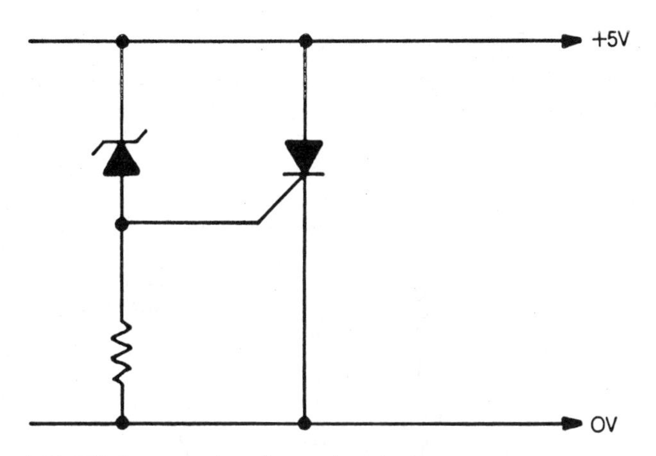

Fig. 6-23. This is one version of a crowbar circuit.

sense circuit for the crowbar. The zener diode will not conduct unless the power supply voltage rises above a predetermined level. So there is no gate current and the SCR does not conduct during normal power supply operation.

If the supply voltage rises above the zener diode conduction level, current follows through it. The resulting gate current turns the SCR on. This places a short circuit across the power supply output. The high cathode-to-anode SCR current causes the fuse to open and the power supply is shut down.

Why not just use the fuse and eliminate the crowbar circuit? The answer is in the speed of the crowbar circuit. It is measured in microseconds. No fuse can respond to an overvoltage that quickly. Also, the crowbar responds to an over*voltage.* Fuses respond to an overcurrent condition. There are applications where an overvoltage can destroy electronic equipment. One example is in some logic and microprocessor systems where a +5V supply must not be exceeded.

Voltage Dividers. Figure 6-24 shows a voltage divider used at the output of a power supply. It would be unusual for this kind of divider to be used with a regulated supply. The reason it is here is that current for the 5-volt output must flow through R1. That makes R1 act as an internal resistance of the supply and ruins the possibility of good regulation. However, you will see this kind of voltage divider used with nonregulated supplies with constant loads.

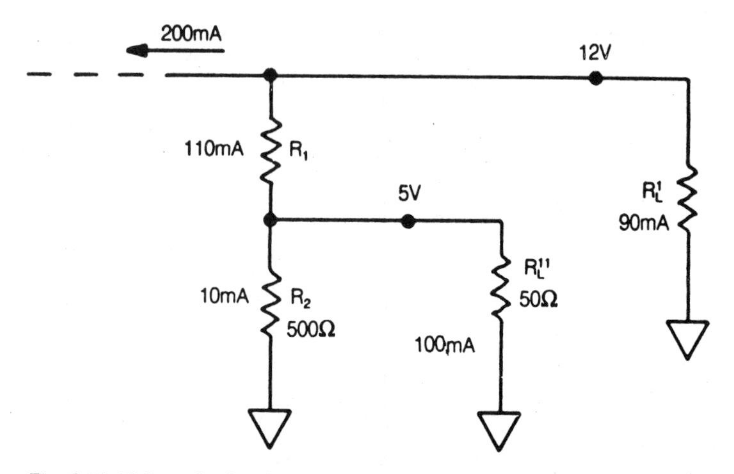

Fig. 6-24. Voltage dividers like this one are used with brute force power supplies.

Calculation of the resistance values depends upon your understanding of Ohm's law and Kirkoff's laws. Consider, for example, the resistance of R1. Here you have two currents combining to flow through R1. They are the 10 milliamps through R2 and the 100 milliamps through R''_L. With 110 milliamps flowing through R1 and a voltage of 7 volts across R1 (12 volts minus 5 volts = 7 volts), the value of R1 can be calculated by Ohm's law as follows:

$$R_l = V/I = 7/0.1 = 70 \text{ ohms}$$

Three-legged Regulators. Integrated circuits are very often used in modern supplies for the regulation circuitry. All of the required circuitry can be placed in a single integrated circuit, which is more efficient from the standpoint space and power drain. These are very popular circuits. The high reliability available from integrated circuits makes the replacement cost of the IC negligible.

An example of a three-legged regulator circuit is shown in Fig. 6-25. You can assume that if the correct unregulated dc is delivered to this regulator, but there is no proper output voltage, the regulator needs replacement.

One of the important characteristics of this type of regulator is built-in current foldback regulation. Current foldback

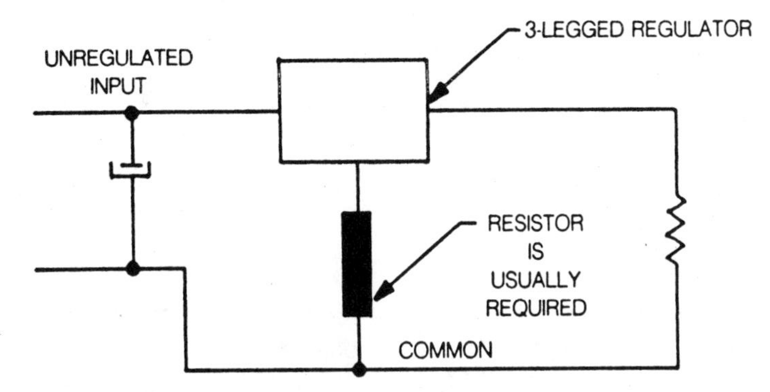

Fig. 6-25. A three-legged regulator saves space and often has special features not found in typical discrete regulators. An example is current foldback regulation.

regulation means that if the output power goes beyond a certain predetermined value, the current and voltage begin to decrease rapidly at the output terminal.

The graph in Fig. 6-26 shows the effect of current foldback regulation. Up to the point where the output voltage or current is excessive, the foldback circuit is transparent to the operation of the system.

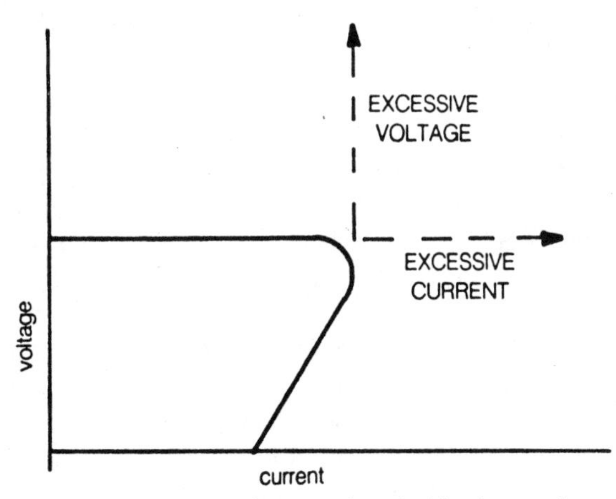

Fig. 6-26. Foldback current regulation is described by this graph.

Filters. Filters in today's systems can be very complicated. Two very simple filters are shown in Fig. 6-27. The first is called a choke input filter. This type of power supply filter

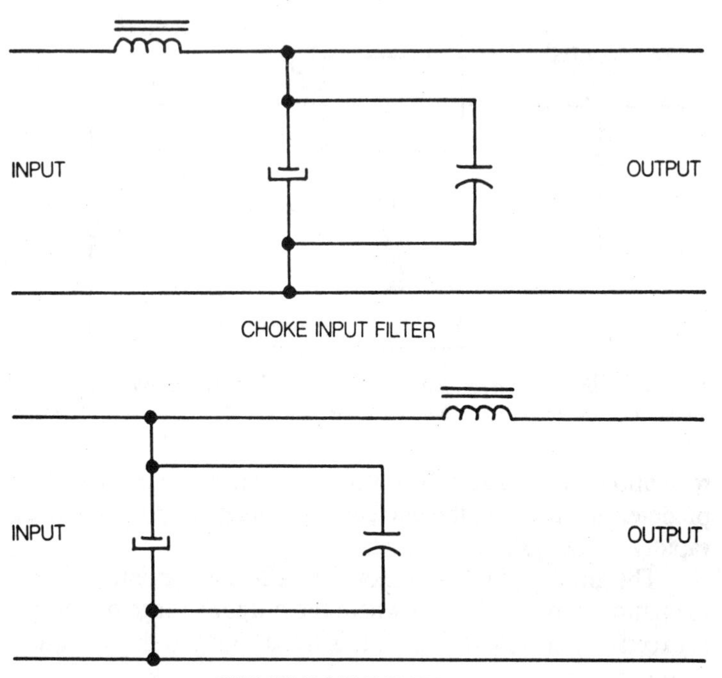

Fig. 6-27. Choke and capacitor input filters are shown here. Stiff voltage regulators do not require this much filtering.

produces better regulation than a capacitive input filter, but it has a lower dc output. In the same illustration the capacitive input filter produces a higher output voltage but it does not produce the good regulation characteristic of choke input filters.

The small capacitors in parallel with the electrolytics serve an important function. Because of the way they are made, electrolytic capacitors have self-inductance. At power line frequencies that inductance is not important. However, if a transient voltage, such as a voltage spike, gets through the power supply, electrolytic capacitors cannot remove it because of their self-inductance.

The small parallel capacitors are for removing the transients. Those capacitors must be constructed in a parallel plane configuration. In other words, they must be nonelectrolytic type capacitors.

SUMMARY

The power supply in an electronic system does not actually supply the power. The power comes from the ac power line delivered by the power company. What a power supply really does is convert the ac to dc. This involves rectifiers, filters, and in many cases a voltage regulator.

Rectifier diodes may be classified by the amount of forward current they have to deliver and the amount of reverse voltage they must be able to withstand. If two diodes are connected in series to obtain a higher peak inverse voltage rating, it is possible that their junction capacitance can be unevenly distributed. That would cause one of the diodes to be destroyed because its peak inverse rating is exceeded. To get around that problem, capacitors are connected across the diodes. These capacitors evenly distribute the reverse voltage. They are chosen to have equal capacitance values within a very close tolerances.

The forward resistance and reverse resistance of the diode is of interest. When a reverse voltage is placed across the diode, some leakage will occur. The leakage resistance of diodes can be unevenly matched. When diodes are in series, that means the highest resistance will produce the greatest reverse voltage. That voltage can exceed the diode's rating, so resistors are connected in parallel with diodes to equalize the reverse voltage drops. Diodes are sometimes connected in parallel to get a higher forward current rating. When that is done, it is necessary to put low-resistance starter resistors in series with the diodes to assure that there is no current hogging.

In a simple half-wave rectifier circuit, the initial charge of the input filter capacitor can destroy the diode. That charging current is measured in amperes. Even though it occurs for a very short period of time, it may have a destructive effect on the diodes.

To get around this problem is sometimes necessary to use surge-limiting resistors. They have a low resistance value and do not seriously affect the output voltage of the power supply in normal operations.

Half-wave and full-wave voltage doublers are used to produce a higher output dc voltage for a given input ac voltage. These circuits use an identical number of components. However, the common connection of the full-wave doubler cannot be connected directly with the ground of the input ac signal. In cases where that is important, the half-wave doubler is used. The full-wave doubler has the advantage of higher ripple frequency, so its output voltage is easier to filter.

Regulators are divided into two classifications: series and shunt. The shunt regulator produces a bypass current that is combined with the load current. The series regulator is set up so the load current must flow through the regulating device. Combinations of series and shunt regulators are often used. An example is the analog regulator.

A voltage regulator can also be classified as being analog or switching. In the analog regulator there is a continuous feedback and adjustment of the output voltage. This can be used to produce a very stiff power supply regulator, but it is not as efficient as the switching regulator.

In the switching regulator, the series pass device is switched on and off. The amount of on time is varied to change the output power and the output voltage. The on time versus the total time is called the duty cycle. Therefore, the switching regulator is duty-cycle controlled, it is sometimes referred to as a pulse width modulation system.

Three-legged regulators are very convenient. They have some special features that are not obtainable with discrete regulators at low cost. One example is foldback current regulating.

Filter capacitors are made by rolling the foil electrodes. This produces an inductance as well as capacitance. In some circuits electrolytic filter capacitors are bypassed by a small planar capacitor. The reason for this is that the inductance of the filter capacitor prevents it from taking out transient voltages. Planer capacitors do not have inductance.

SELF TEST

1. Which of the following will produce the largest depletion region?
   (A) a forward biased semiconductor rectifier diode.
   (B) a reverse biased semiconductor rectifier diode.

2. When two capacitors are in series across a dc voltage, the larger voltage is across the smaller capacitance value.

When they are across an ac voltage source, the largest voltage drop is across the:
(A) larger capacitance value.
(B) lower capacitance value.

3. Which of the following PN junctions has the higher forward voltage drop during conduction?
   (A) silicon.
   (B) gallium arsenide.

4. What is an advantage of the full-wave doubler over the half-wave doubler?
   (A) It is easier to filter the output.
   (B) It uses fewer components.

5. A disadvantage of the full-wave doubler is that it:
   (A) does not have as high an output voltage as the half-wave doubler.
   (B) does not have a common connection that can be directly connected to the ac ground for safety.

6. A zener diode regulator is an example of a:
   (A) series regulator.
   (B) shunt regulator.

7. In an analog regulator, an operational amplifier is used as a:
   (A) sense amplifier.
   (B) reference amplifier.

8. Which of the following is correct?
   (A) Duty Cycle = On Time/Off Time.
   (B) Duty Cycle = On Time/Total Time.

9. In a switching regulator, pulses are delivered to the series pass amplifier by:
   (A) an oscillator.
   (B) a sense amplifier.

10. Which of the following is correct?
    (A) Voltage dividers are most often used with a stiffly-regulated supply.
    (B) Voltage dividers are usually used with a non-regulated supply.

ANSWERS TO SELF TEST

1. (B)

2. (B)

3. (B)

4. (A)

5. (B)

6. (B)

7. (A) This op amp is often called a difference amplifier as in Fig. 6-18. However, it really amplifies the sense signal and the reference circuit provides a dc value for comparison with the sense input.

8. (B)

9. (A) The relaxation oscillator used to deliver the pulse to the series pass regulator can be any of the relaxation types, such as the multivibrator or an oscillator produced by an integrated circuit timing device.

10. (B)

7

Motors, Generators, and Control

YOU DO NOT have to be an expert in motors and generators to work in industrial electronics. However, remember that much industrial electronic circuitry is used for controlling motor speeds and generator voltages. For that reason you should have a good basic understanding of how motors and generators work because it will better help you to understand their control.

Both dc and ac motors are used in industry and each has their special advantages and disadvantages. DC motors are very popular because their direction can easily be reversed and it is a simple matter to control their speed, especially at very low speeds.

By contrast, ac motors do not have a high torque development at very low speeds, so it is common practice to use some type of belt and pulley or transmission system to get satisfactory torque at low speed.

DC motors have a very high torque at start-up speeds. This is especially true of a series motor discussed in this chapter. For portable operations, dc motors are preferred because they can be operated directly from a battery. Also, generators or alternators can be used to charge a battery when it is away from any ac source. When near an ac source, a rectifier power supply can be used to provide the dc necessary for charging batteries.

The biggest disadvantage of dc motors in industry is that dc is not readily available. When dc motors are used in industrial processes, it is necessary to deliver the dc from a central source by using very large bus bars. That is expensive compared to delivering ac power to an ac motor.

AC motors, then, have the advantage that the power for operating them is readily available. These motors are economical and relatively simple to construct. They do not usually require brushes and commutators. Therefore, they require very little maintenance during their lifetime of operation.

By contract, dc motors must have their brushes periodically replaced. Failure to do that can be destructive to the commutator and that means additional maintenance problems.

There are certain types of ac motors that have a built-in speed control. As you will see in this chapter some motors have their speed directly controlled or dependent upon the frequency of the input signal.

CHAPTER OBJECTIVES

After studying this chapter, you will be able to answer the following questions:

- What is the basic construction of dc motors and generators?
- What is a series-wound dc motor and what are its advantages and disadvantages?
- How can a dc motor be used as a generator and how is this capability used to stop the dc motor?
- Why are capacitors needed for starting some types of ac motors?
- Which type of ac has a speed determined by the input ac frequency?

PHYSICAL LAWS AND EFFECTS RELATED TO ELECTROMECHANICAL DEVICES

There are some basic physical laws and effects that are important in understanding the operation of many electromagnetic devices. Some examples of those devices are:

- Transformers
- Motors
- Generators
- Relays
- Solenoids
- Hall devices
- Ferrite beads

Faraday's Law. Faraday's law states that any time there is relative motion between a conductor and a magnetic field, a voltage is induced in the conductor. Refer to Fig. 7-1. Here a conductor is positioned between two magnetic poles. If that conductor is moved vertically, a voltage will be induced across its ends because it is moving through the magnetic flux between the poles. If a conductor is moved horizontally, there is no voltage induced because it is not cutting across any flux lines.

The distribution of flux lines between the poles is also shown in Fig. 7-1. If the conductor is moved at an angle to the flux, the induced voltage will be less than for vertical motion and greater than for horizontal motion. A greater angle between the motion and the flux lines results in a higher induced voltage.

Faraday's law describes the amount of voltage induced. Mathematically, this equation is written as follows:

$$v = -N\,(d\phi/dt)$$

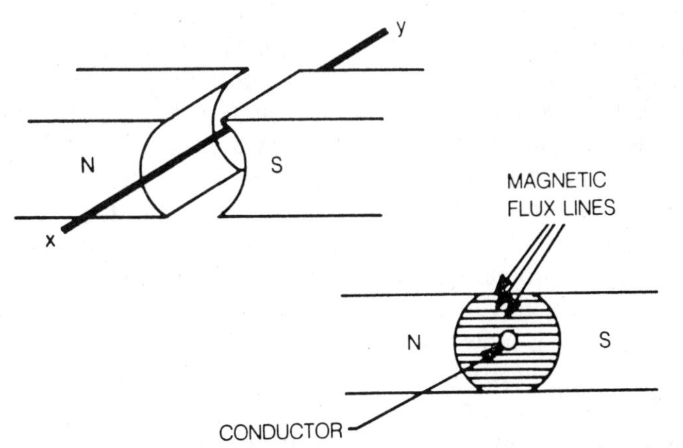

Fig. 7-1. Conductor XY is suspended between the magnetic poles. The side view shows the distribution of flux between the poles.

where N is the number of conductors moved through the magnetic field, and $d\phi/dt$ is the rate at which the conductor is cutting through the flux lines.

The higher the rate of cutting flux lines, the higher the value of $d\phi/dt$. You can see from the equation that you can increase the induced voltage by moving more than one conductor at the same time and connecting them so that their voltages add. Likewise, you can increase the voltage induced by increasing the number of turns.

Lenz' Law. The negative sign in the equation for Faraday's law is used to indicate that the induced voltage has a certain polarity. Specifically, if a connection is made across the ends of the conductor being moved, there will be a current flow. In this book we will define that current as being an induced current.

The induced current, like all currents, has a surrounding magnetic field. That magnetic field will always be in such a direction that it opposes the motion that produced it. This is a very important concept. It indicates that you have to do work (force × distance) to move the conductor through the magnetic field if that conductor is part of a closed current path.

Voltage is a unit of work. Technically, voltage is the amount of work done in moving a unit charge of electricity around a closed path. The amount of voltage generated is related to the amount of work done in generating it. If it is part of a closed-loop circuit, the amount of work required is greater than if the conductor is not connected. That is why a generator is harder to turn when it is delivering current.

Lenz' law is responsible for the countervoltage induced in a conductor when it is moved through a magnetic field. This assumes that a conductor has a closed loop and there is an induced current. The countervoltage generated in the conductor always opposes the induced current and it opposes the motion that produced it. So, if a current is flowing from left to right, the countervoltage is from right to left.

Figure 7-2 shows a method of determining the polarity of an induced voltage when a conductor is moved through a magnetic field. Remember that the direction of magnetic flux is always away from the north magnetic pole and toward the south magnetic pole.

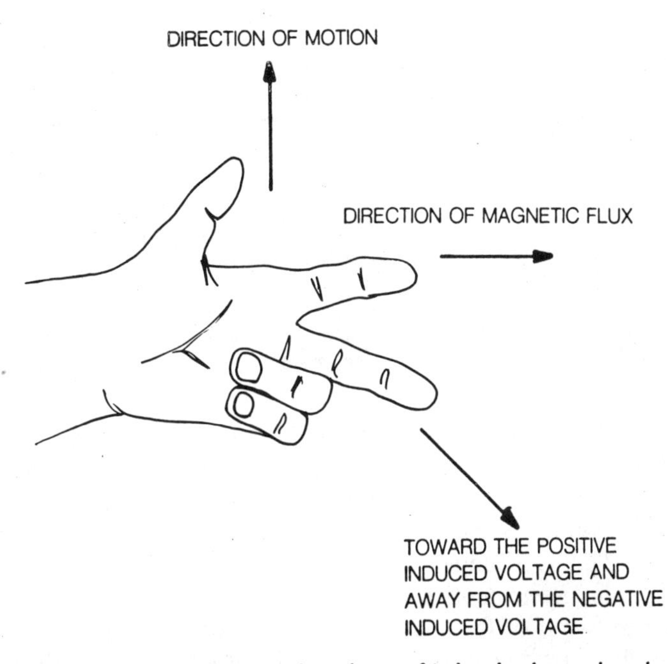

Fig. 7-2. The left hand rule shows the polarity of induced voltage when the conductor is moved in the direction indicated by the thumb. The hand is presumed to be inside the generator.

To summarize, voltage is induced when a conductor moves through a magnetic field. The voltage is directly related to the number of conductors and the speed that the conductors move relative to the flux line. Whenever voltage is induced, an induced current will flow, provided the ends of the conductor are connected in a closed circuit. The induced current always has a magnetic field that opposes the motion of the conductor. Furthermore, when the conductor moves and there is a closed circuit, there is a countervoltage that opposes the induced current. A countervoltage will always oppose any change in the magnitude of the current in an inductive circuit.

In Fig. 7-3 the conductor is being moved through the magnetic field and there is a complete path for current flow. Observe that the electron current flow is from − to + . If you are using conventional current flow, the direction of current will be reversed. However, the polarities of the induced voltage will be the same.

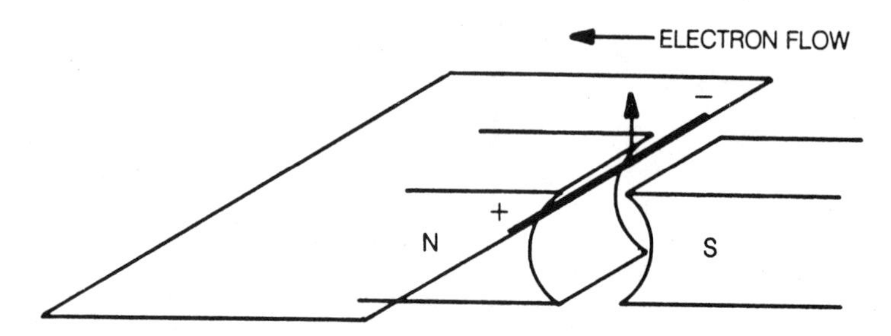

Fig. 7-3. In this illustration, a conductor is connected between the ends of the generated voltage. Note that electron flow is from − to + outside of the generator, but it is from + to − inside the generator.

You should take the time to apply the left-hand rule to verify that the voltage polarities in Fig. 7-3 are correct based on the direction of motion of the conductor and the north-to-south direction of the magnetic field.

Countervoltage. At one time countervoltage was called counter electromotive force. However, the term *electromotive force* is out of favor. It has caused a considerable number of problems with students and technicians who went to advanced work. Voltage is a unit of work. It is not a unit of force.

This is not an important point at the technician level, but in advanced work it becomes necessary to work with equations and to rationalize units. It is never possible to do that and come out with voltage as a unit of force. It always comes out as a unit of work.

Nevertheless, considering voltage a force that pushes electrons through the circuit is a very good model as long as you remember that it is only a model.

Figure 7-4 shows two conductors being moved simultaneously through a magnetic field. The original conductor (XY)

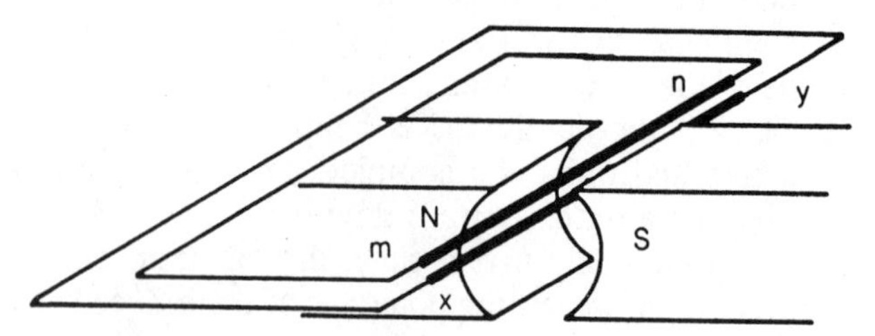

Fig. 7-4. Parallel conductors will result in a countervoltage.

is now in parallel with the new conductor (MN). There are complete paths for current to flow, so there are induced currents in both lines.

You will remember that whenever a current flows in a circuit, there is always an accompanying magnetic field. Therefore, the magnetic field for MN will cut across XY and the magnetic field from XY will cut across MN. Therefore, each conductor will induce a countervoltage in the parallel conductor inside the generator. That countervoltage will always be in such a direction that it opposes the flow of induced current in the external circuit.

In a typical motor or generator there is a large number of parallel conductors—each generating voltages in nearby conductors. So, every motor has a countervoltage generated inside it. Also, every generator has a countervoltage generated inside it.

Ampere's Law. One important concept that has been mentioned should be discussed further. It has been stated that whenever there is a current flow, there is always an accompanying magnetic field. That magnetic field circles the current. According to Ampere's law, the strength of the magnetic field is directly proportional to the amount of current flowing.

The left-hand rule is sometimes used to define the direction of the magnetic flux around an electron-current carrying conductor. If you (mentally!) grasp the conductor with your left hand so that your thumb points in the direction of electron flow, your fingers will circle the conductor in the direction of the accompanying magnetic field. That is always north-to-south.

Again, for conventional current flow you will use the right hand, pointing your thumb in the direction of conventional current flow. Your fingers will still grasp the wire in the direction of the magnetic flux.

The Motor Rule. So far we have discussed the generation of voltages and countervoltages when conductors are moved through a magnetic field. Now consider what happens when a current-carrying conductor is placed in a magnetic field. This is the principle upon which motors are based.

The magnetic field accompanying a current interacts with

the magnetic field between the poles of the magnet. If the conductor is free to move it will do so.

The direction of motion of the conductor can be determined by the right hand motor rule. As with the left hand generator rule, this is another form of Flemming's rules. Flemming's rules were originally based upon conventional current flow.

Figure 7-5 shows how the right hand rule applies. Again, the forefinger points in the direction of magnetic flux. The electron current through the conductor is represented by the second finger. The thumb now points in the direction the conductor will move if free to do so. You can see that the conductor will move up, in the illustration of Fig. 7-6.

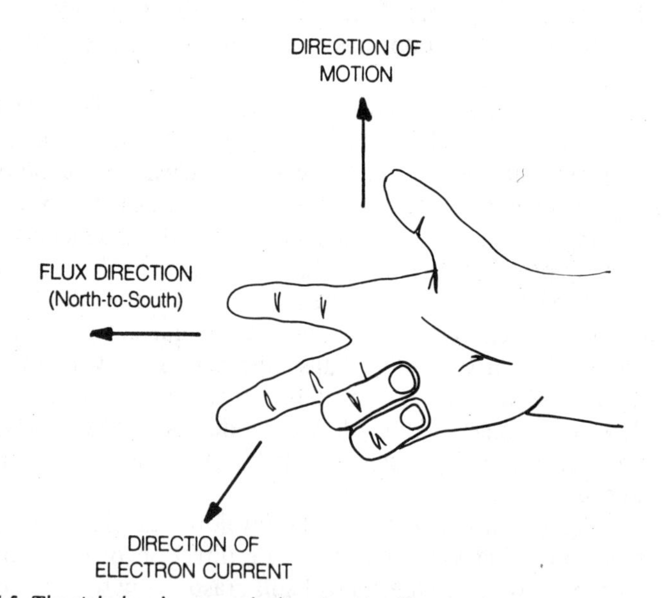

Fig. 7-5. The right-hand motor rule describes the direction that a current-carrying conductor will move when placed in a magnetic field.

Figure 7-7 shows two important models of current-carrying conductors in magnetic fields. To understand this model you should understand that magnetic flux lines always try to take the shortest possible distance. One concept compares them to rubber bands. You can stretch them out of their normal posi-

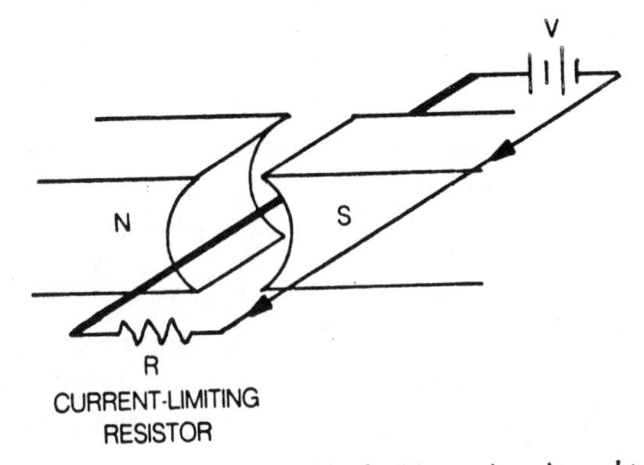

Fig. 7-6. An external power supply and a limiting resistor is used to move the conductor.

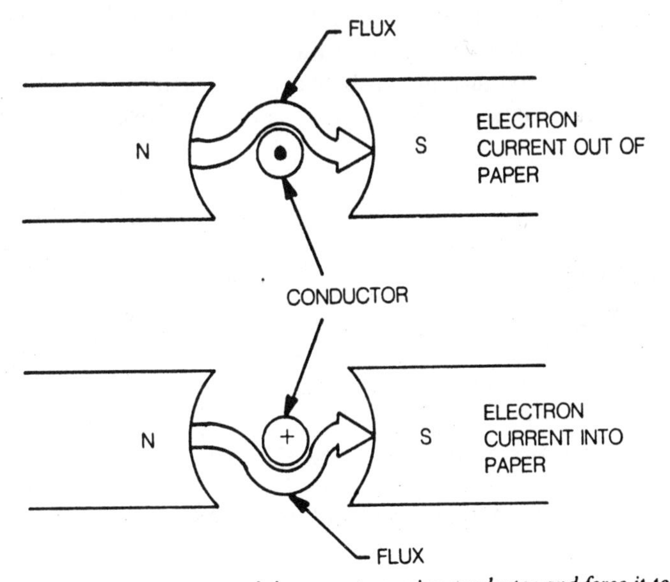

Fig. 7-7. Flux lines wrap around the current-carrying conductor and force it to move.

tion but they will always try to return to the position of least distance.

The symbolism in Fig. 7-7 is standard. A dot in the center of the conductor means that the electron current is moving toward you. (Think of the tip of an arrow.) The plus sign in the conductor means that the electron current is moving away from you. (Think of the tail of an arrow.)

Keep in mind that there is a magnetic field surrounding the conductor when it is carrying a current. The flux of the permanent magnet will push against the magnetic field around the conductor and force it to move. In the upper illustration of Fig. 7-7, the conductor is forced downward. Use the right hand motor rule to verify this. In the lower illustration, again using the right hand rule, the conductor is pushed upward.

Keep this illustration in mind because the distortion of the magnetic field becomes an important thing in practical motors and generators.

The Left-Hand Rule for Coils. Before leaving the subject of physical laws and effects, there is one more rule that you should understand. It is illustrated in Fig. 7-8. A current carrying coil is wrapped around a piece of soft iron material. Soft iron is a material that can be easily magnetized, but does not retain its magnetism when the magnetizing force is removed.

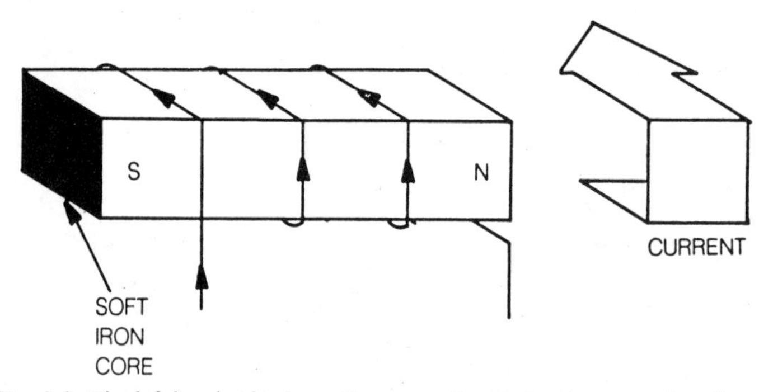

Fig. 7-8. The left-hand rule shows the magnetic polarity for a simple coil.

The direction of the magnetic field induced in the soft iron can be determined from the direction of the current in the coil. A left hand rule is used in this case. Grasp the coil (in your imagination only!) with your fingers pointing in the direction of electron current around the coil and your thumb will then point in the direction of the north pole of the induced magnetism.

For conventional current flow, the right hand rule is used. The electron current would then be reversed. The coil would be grasped with the right hand and the thumb would point in the direction of the north pole.

You may well wonder about the use of some of the physical laws and effects that were discussed in this chapter. Does an industrial electronics technician really need to know these things?

Actually, the rules, physical laws and the effects are primarily for improving your understanding of motors and generators. They are not necessarily things that you would apply in your job every day. This is not only true of the physical laws and effects in this chapter. In all chapters and in all books, technical theory is presented to help your understanding of the subject. Not every single sentence is directed to things you can do on your job. Instead, authors want to make it possible for you to have a better understanding of how the things work. If you know how things work, you will be better able to find out what is wrong when they are not working. From this point we will discuss some practical motor and generator ideas.

MORE PRACTICAL GENERATORS AND MOTORS

Consider now the illustrations in Fig. 7-9. Instead of being straight, the conductor has been bent into the shape of a loop. We will first consider what happens when this loop is turned in the magnetic field.

Using the rules given for generators, you will see that with a clockwise motion, the voltage induced in one side of the loop adds to the voltage in the opposite side of the loop. Therefore, twice as much voltage will be generated when the loop is turned in the magnetic field.

In the zero position, the loop is cutting through the maximum number of flux lines per instant of time. Therefore, the induced voltage is maximum. After the loop has been turned 90 degrees, the conductors of the loop are moving parallel to the flux lines between the poles. Since they are not cutting across conductors, there is no voltage being generated.

After the loop has passed the 90 degree position, the conductors again move through the magnetic field. If you keep track of the direction of the induced voltage in one side of the

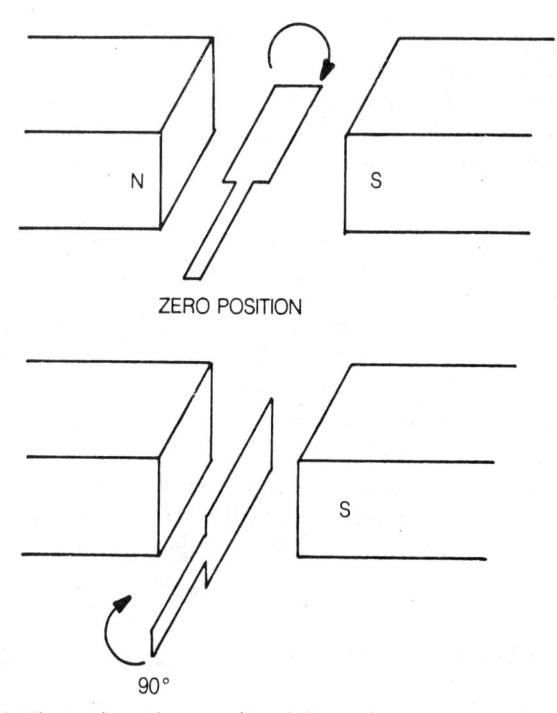

Fig. 7-9. The conductor between the poles is made into a loop. Two positions for the loop are discussed in the chapter.

loop you will see that it reverses after the loop passes the 90 degree point.

If the loop is turned at a constant speed you will get an ac voltage across the ends of the loop. That would be OK for an ac generator but since we want dc current here, the output of the loop will have to be rectified in some way. Of course, diodes could be used. In that case the device becomes an alternator, which is a dc generator obtained by an ac generator with a diode rectifier connected to its output. The alternator in cars is based upon this principle of operation.

Normally, the generator output is rectified by a mechanical rectifier. The mechanical rectifier is made of a commutator and carbon brushes.

The loop of wire of Fig. 7-10 is physically and electrically connected to a commutator. As the loop rotates, so does the commutator. In the positions shown, the maximum voltage is

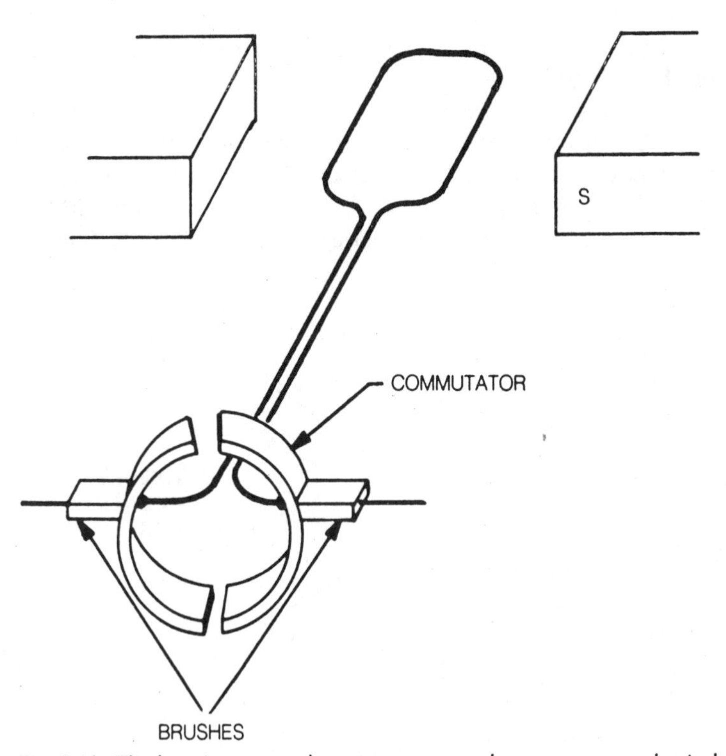

Fig. 7-10. The loop is connected to a commutator that acts as a mechanical rectifier.

being delivered to the commutator segments so, maximum output is obtained from the brushes.

If you mentally rotate the loop and commutator 90 degrees, while holding the brushes in their position, you will see that the brushes are connected to the slots in the commutator ring. In practice these slots are filled with a material that keeps the brushes in position. Brushes are pushed against the commutator segments by springs in order to get good contact.

As the loop continues to turn, the section that was originally on the left side in Fig. 7-10 is now on the right side. However, the commutator has switched this side of the loop to the right-hand brush. Likewise, the loop that was on the right-hand side is now on the left-hand side and the commutator has switched this side of the loop to the left side of the brush. The overall result is that the output voltages will have the same polarity as before. The voltage has been mechanically rectified by the commutator.

In the simple generator in Fig. 7-10 the commutator produces a half-wave rectified voltage. In practice the commutator has many more segments corresponding to many more loops. This idea is represented in Fig. 7-11. As shown in the illustration, there are also more magnetic poles and more conductors.

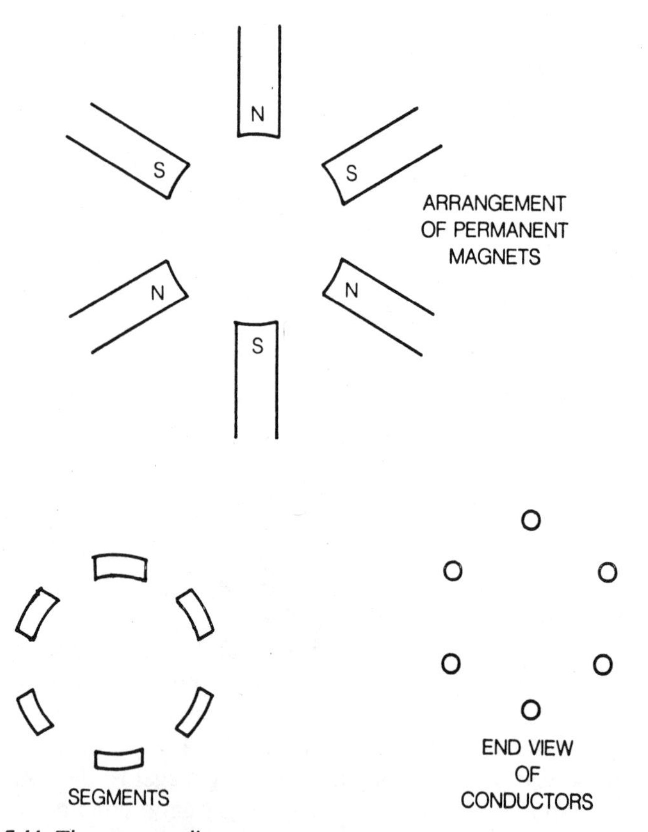

Fig. 7-11. There are usually more commutator segments, more poles, and more loops. This gives a smoother operation.

The overall result is that the output voltage at the brushes never actually drops to zero volts. The result is a dc voltage with a very slight ripple, dependent upon the number of commutator segments and loops.

Interpoles. The illustration in Fig. 7-7 has been upgraded in Fig. 7-12. The two ends of the loop are being rotated in the magnetic field. The distortions of the flux lines, due to each half of the loop, are illustrated.

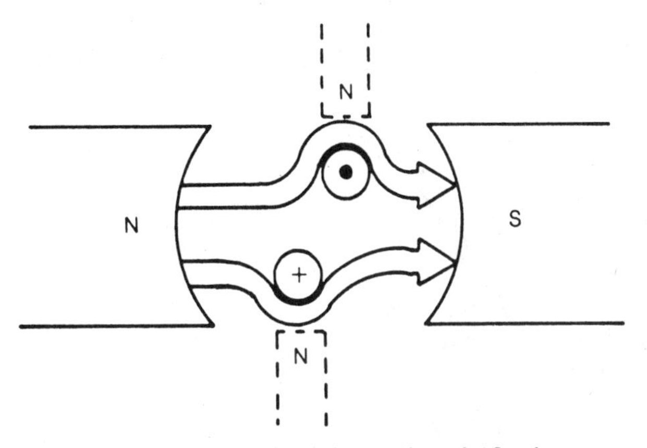

Fig. 7-12. Interpoles may be used to help straighten the flux lines.

This distortion of flux lines occurs in both motors and generators. In some cases interpoles are used, as shown with broken lines, to reduce the effects of flux distortion. In other words, the interpoles straighten the flux lines. If the flux remains in its normal position, there will be a greater force or greater voltage on the conductors when there are interpoles. These interpoles are normally connected inside the generator.

Instead of turning the loop between the magnetic poles to produce a generated voltage, it is possible to apply a voltage through the commutator segments. That is how dc motors are made. The basic concept is illustrated in Fig. 7-13.

The rotating loop and commutator segments are mechanically fixed to the shaft of the motor. As the loop and segments turn, the motor shaft turns. The motor shaft is not shown. For the position shown in Fig. 7-13, the maximum turning force—or *torque*—will be produced.

When the loop has turned 90 degrees, the segments in the commutator reverse the currents in the two loop halves. The motor continues to turn.

In practice, a motor of the type in Fig. 7-13 will work, but it does not have the smooth operation usually required for dc

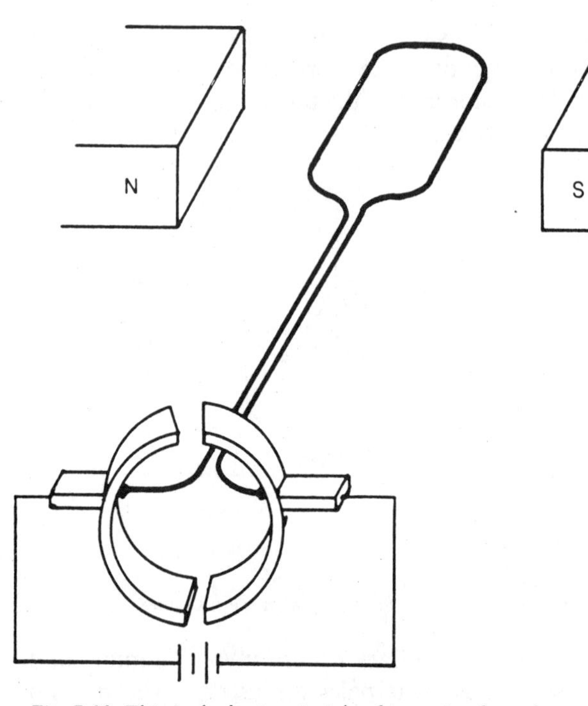

Fig. 7-13. This is the basic principle of operation for a dc motor.

motors. So, the motor is modified as shown in Fig. 7-11 to include a number of magnetic poles, conductors, and commutator segments.

DC Motor-Generator Action. From the similarity in the illustrations in this chapter, you can see that dc motors and generators have the same basic construction. A permanent magnet dc motor will produce a voltage at its terminals if its shaft is rotated. Conversely, the permanent magnet dc generator will run like a motor if a voltage is applied to its terminal.

This concept is sometimes used to test generators. A dc voltage is applied to see if a generator will rotate like a motor.

Likewise, motors can be checked to see if they will produce a dc output when their shaft is rotated. These are just quick checks in a troubleshooting procedure. They do not tell you anything about the condition of the motor or generator.

The similarities between motors and generators are also used to other advantages. Suppose, for example, a dc motor is turning at high speed and it becomes necessary to stop it quickly. If you short its terminals together, after removing the applied voltage, a countervoltage is induced in the conductors and an induced current flows that opposes the rotation. This procedure is called dynamic braking. A motor can be stopped very quickly using this technique.

Another method for stopping a dc motor quickly is to reverse the polarity of voltage applied to its terminals. This method is called plugging.

In some battery-operated equipment the technique is slightly modified. When the motor is running at full speed, it is doing its normal work. When the motor is stopped, the countervoltage is used to charge the battery, thus insuring a longer time interval between major charges.

Motor/Generator Excitation. A magnetic field is absolutely necessary for the operation of these types of generators and motors. In dc applications the magnetic field is constant. It may be produced in any of three ways.

First, it can be produced by a permanent magnet. In today's technology, permanent magnets can produce very strong fields and they do not deteriorate with time as in earlier days.

The second method of producing a magnetic field is called self-excitation. When a generator is self-excited, it means that the dc produced by the generator is not only delivered to the external terminals but is also used for producing the magnetic flux. This situation is illustrated in Fig. 7-14.

With the switches in the position shown, the output of the generator commutator delivers a current through the field coil. Since this current is in series with the armature, this is a series-wound generator. The important point is that the coil current is produced by the generator itself. This same technique can be used for shunt and compound-wound generators.

The third method of producing the magnetic flux occurs when the switches are set opposite from what is shown in Fig. 7-14. Now, an external power supply (V) is used to produce current in the coil. This is called a separately excited motor or generator field.

In this illustration, a variable resistor limits the amount of current through the coil. When the motor is first started, the coil

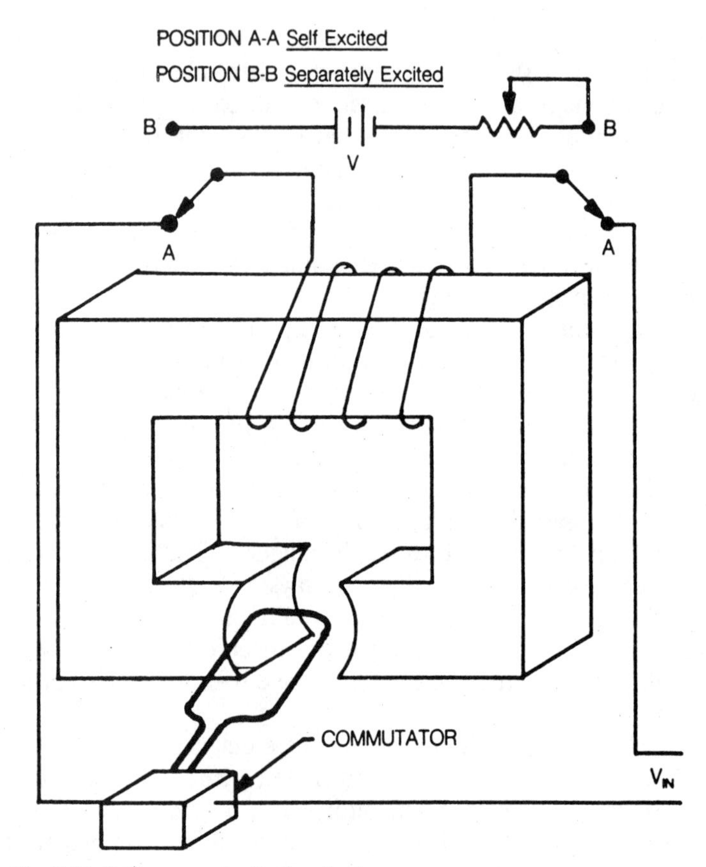

Fig. 7-14. Different methods of excitation.

current must be very low because they is very little opposition to current flow other than the resistance of the coil and the resistance of the rheostat in series with the coil.

When the motor or generator is operating at full speed, a countervoltage is generated in its coils. That countervoltage opposes the current through the field coil, so the resistance in series with the coil of Fig. 7-14 must be decreased in order to make full use of this separate excitation.

If the variable resistor is set in one position, it is not possible to get a good compromise between the maximum possible current in the coil when the motor or generator is running and the minimum current when it is first starting. Therefore, the variable resistor must be operated manually as the motor or generator is brought up to speed.

A variable resistor is not an efficient way to do this. Instead, it is a common practice to use manual starters to bring the motor or generator up to its full capability. This is primarily applied to motors, but theoretically it could also be used for generators.

Operation of a Manual Starter. The manual starter in Fig. 7-15 is one example of how manual circuits can be used to start a motor and bring it up to its maximum speed. Electronic starters are also available that accomplish the same thing.

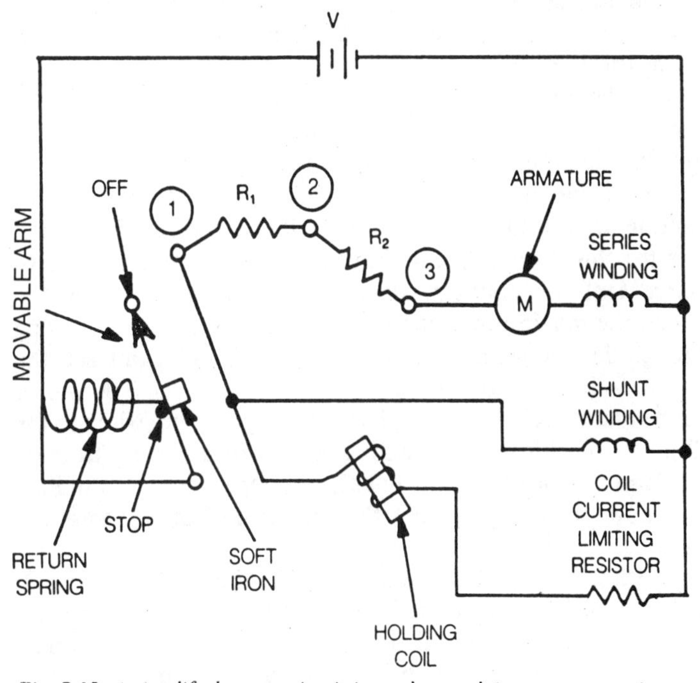

Fig. 7-15. A simplified starter circuit is used to explain starter operation.

When a motor first starts to turn, there is very little opposition to the current flow through the armature because there is no countervoltage. As the motor speed builds, the armature countervoltage develops and increases. This countervoltage opposes the supply voltage and power to the motor. It is necessary to provide some limitation to armature current during startup. The most common method of doing this is to use a manual starter.

We will describe a simple manual starter to establish the principle of operation. Refer to Fig. 7-15. The arrow represents a manually turned switch. It is sitting against the stop in the off position. A return spring provides tension to hold it against the stop. Notice that there is a soft iron cube mounted on the manual switch.

When the switch is turned to position 1, current flows through the shunt winding. It also flows through R1 and R2, through the armature, and through the series winding. The motor begins to turn and the armature current is limited by the two resistors in series.

As the motor speed increases, the operator switches to position 2. That reduces the resistance in series with the armature, increases the current, and the motor turns faster.

When the motor is up to full speed the switch is turned to position 3. Now the armature and series winding are in parallel with the shunt winding and they are directly across the voltage source. Only the countervoltage limits the armature current.

In the maximum speed position there is a current flowing through the holding coil. It has a magnetic field that attracts the soft iron cube and holds the switching position 3. The motor can be manually stopped by physically moving the switch to the off position. This requires pulling it away from the electromagnet. Manual starters usually have more steps than the simplified one shown in Fig. 7-15, but the principle of operation is the same.

Series, Shunt, and Compound Windings. So far, much of the discussion has centered on permanent magnet type generators and motors. Long life permanent magnet materials now make these types of generators and motors very practical. But, the amount of voltage generated and the amount of turning free (or torque) for the motors is limited by the amount of magnetic flux available. As a general rule it is possible to get a greater amount of flux using an electromagnet compared to the flux available in a permanent magnet having the same size.

For a generator, the current for the electromagnet can be obtained from the generator itself in a self-excited device. There are a number of ways to connect the electromagnets. The coils for these electromagnets are called the field coils. The rotating loop between the magnetic poles is called the armature. Figure 7-16 shows the possibilities.

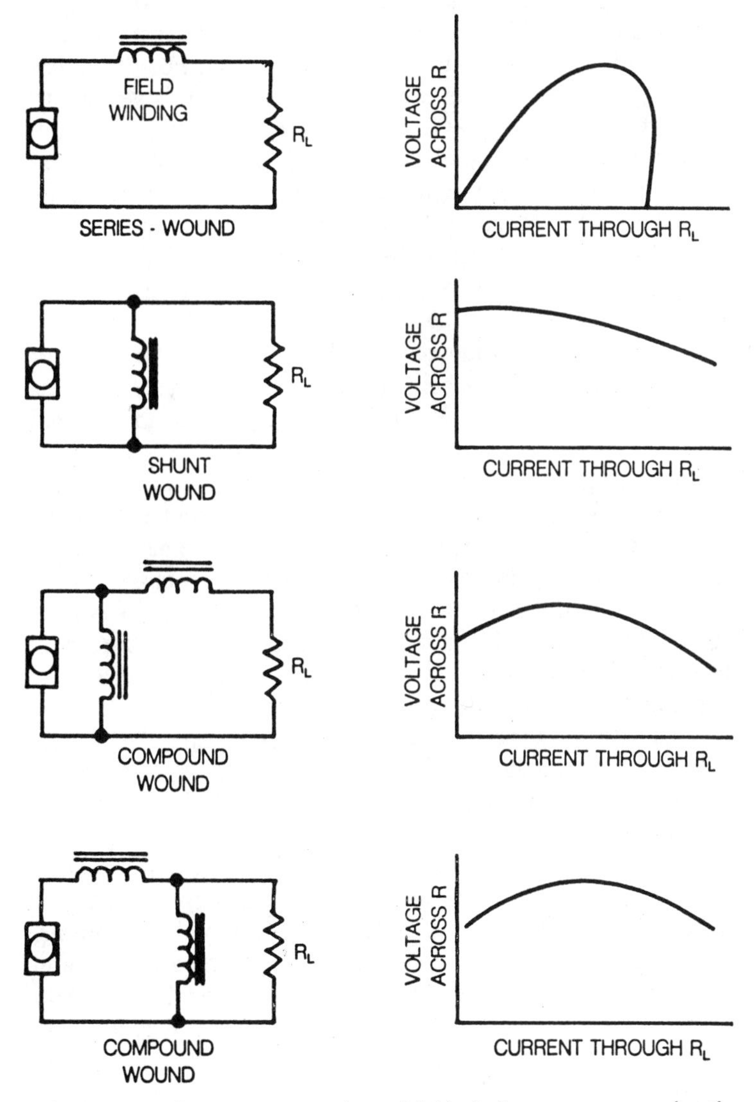

Fig. 7-16. For the various connections of field windings, you can see that the generator output is greatly influenced by the construction of the generator.

If the field coil is connected to the armature in such a way that the same current that flows through the coil also flows through the load resistance, you have a series-wound generator.

You will remember from an earlier discussion on power supplies that regulation is a measure of how well the output of a battery power supply or generator maintains its current and its output voltage for various amounts of current. For the series-wound generator you can see that the regulation is very poor.

In the shunt-wound generator, the field winding is in parallel with the load resistance. This produces much better regulation and this is the way typical self-exciting generators operate.

The next two figures illustrate compound wound connections. By suitably choosing the electromagnetic characteristics of the series and shunt coil, it is possible to get special characteristic curves.

As a technician you would not be able to tell which kind of compound windings are being used and how the generator was designed except that its output does not produce the normal series or shunt generator characteristic curves.

Windings in Practical Motors. Consider again the illustration in Fig. 7-16. If you replace the load resistors with batteries, the device becomes a motor instead of a generator. The same names apply. For example, when the battery current flows through both the armature and the field, it is a series-wound motor. Likewise, if the battery current runs through the armature and field and they are connected in parallel across the battery, it is called a shunt-wound motor.

As with the generators, the method of winding the electromagnetic field connecting it with the armature has an important influence on the operation of the motor. This is illustrated in Fig. 7-17.

Consider, again the illustration in Fig. 7-16. If you replace the load resistors with batteries, the device becomes a motor instead of a generator. The same names apply. For example, when the battery current flows through both the armature and the field, it is a series-wound motor. Likewise, if the battery current runs through the armature and field and they are connected in parallel across the battery, it is called a shunt-wound motor.

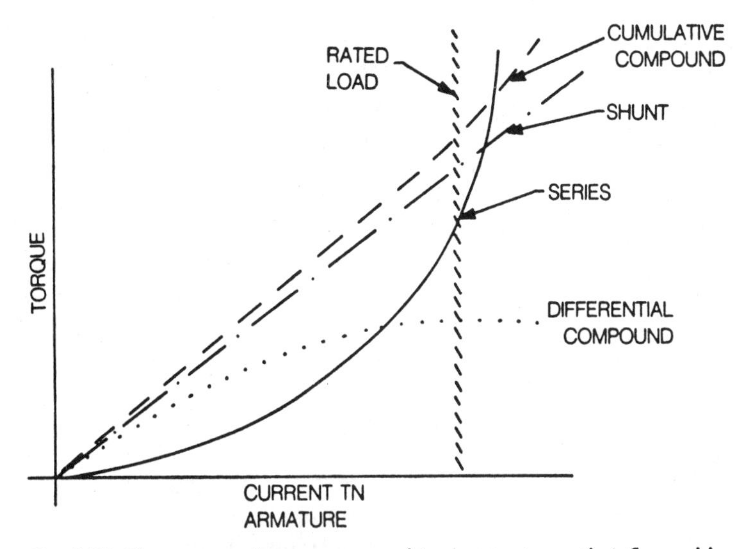

Fig. 7-17. The amount of torque generated in a motor is greatly influenced by the construction of the field windings.

Consider, first, the characteristi urve of the series-wound motor. You can see that as the current increases, the speed and torque of the motor increase very rapidly. Series-wound motors have a very high starting torque. An important problem with them is that if you connect the series-wound motor across a battery without a load, it will continue to go faster and faster until ultimately it will reach a speed at which it will destroy itself. For this reason these motors must never be connected without a mechanical load.

Look at the more nearly constant torque and current characteristics of the shunt-wound motor. This motor will not run itself to death in normal operations without a load. There is one problem, however, you should be aware of. If the field winding opens, there will only be a small amount of magnetic flux left in the soft iron material used for making the field. That small amount of flux will be sufficient to cause the motor to turn. The important thing is that there will be very little countervoltage and very little opposition to the turning. In that case the shunt-wound motor will behave like a series-wound motor. It will run faster and faster until it ultimately destroys itself unless it is connected to a mechanical load.

As with generators, it is possible to get special effects by the way the series and shunt field windings are combined in a dc motor. Again, you probably will not know how the windings are made unless you are a motor technician. As an electronic technician you are primarily interested in what happens if the windings do not work. Remember: if the shunt winding is open, the motor will act like a series wound motor, and if the series winding is open the motor will not run at all.

AC GENERATORS AND MOTORS

AC generators and motors are less complicated than their dc counterparts. They do not require commutators. In most cases they do not have brushes. Remember that brushes are made of carbon—the same material used to make resistors. Therefore, they cause heat dissipation and a voltage drop. Those losses can be avoided in most ac generators and motors.

AC Generators. Figure 7-18 shows a simple method of generating an ac voltage. Here, a permanent magnet is rotated between the field poles. In this case the poles are made of soft iron and are wound with two coils in series.

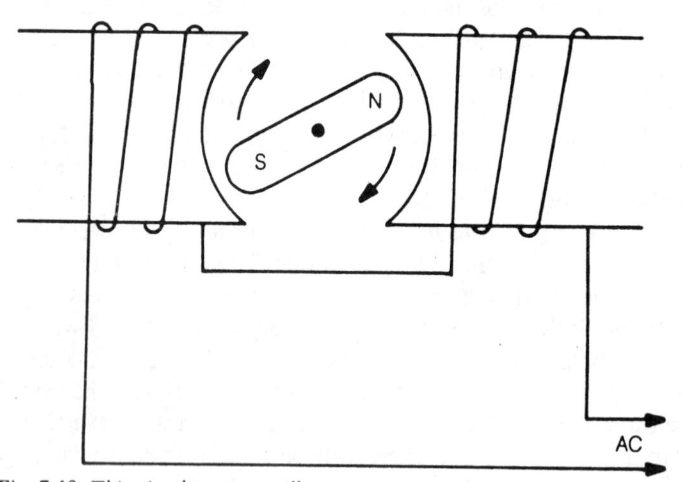

Fig. 7-18. This simple system will generate an ac voltage.

As the magnet turns, the flux induced in the soft iron is first in one direction and then in the opposite direction, depending upon the positions of the permanent magnetic poles. The frequency of the ac will depend upon the speed at which the rotating magnet is turned.

In more sophisticated ac generators, an electromagnet is rotated rather than a permanent magnet. Connection to the internal rotating electromagnet is through slip rings. They are illustrated in Fig. 7-19. Brushes deliver the direct current necessary for the electromagnet.

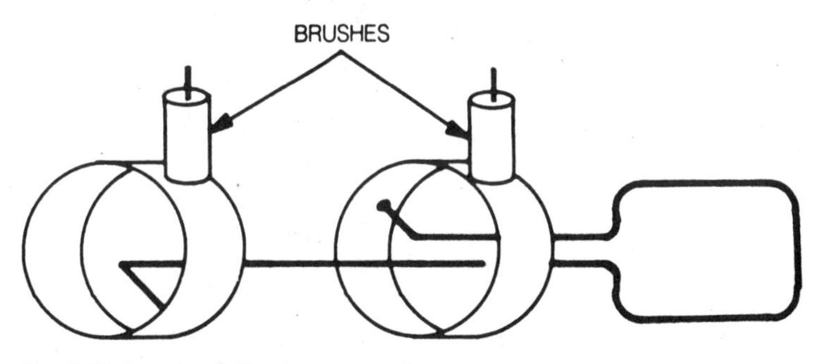

Fig. 7-19. Details of slip-ring construction.

In another type of ac generator, a conducting loop is rotated between north and magnetic south poles and the resulting ac is delivered through slip rings and brushes to the outside world.

As you can see, the construction of an ac generator is quite simple. The most efficient design is the rotating permanent magnet (Fig. 7-18) with the ac taken from the generator without the need for slip rings or brushes.

AC Motors. Many types of ac motors make use of a rotating magnetic field which is followed by a soft iron or permanent magnet core material. In one of the most popular types, a rotating combination of conductors—called a *squirrel cage*—is used as an armature.

Figure 7-20 shows the concept of a rotating field. In the first illustration, the permanent magnet is aligned with block A which is presumed to be a soft iron material. Because of their proximity, the soft iron has an induced magnetic field from the

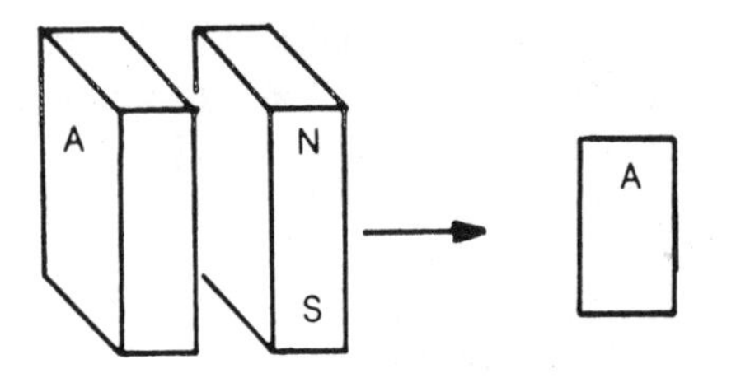

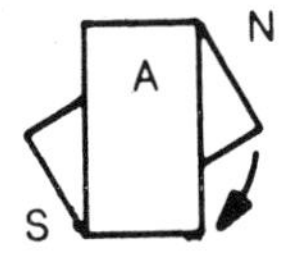

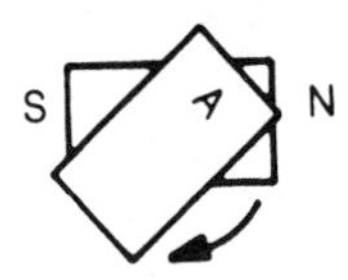

Fig. 7-20. The concept of rotating magnetic field.

permanent magnet. Therefore, both blocks have magnetic fields.

When the permanent magnet starts to turn, the magnetized soft iron will begin to follow it. The shaft of the motor is connected to the rotating soft iron bar.

The Rotating Magnetic Field. There are several ways of getting a rotating magnetic field. One way is to use two-phase power as shown in Fig. 7-21. The waveforms are 90 degrees out of phase. The result is that the north pole of the induced magnetism will rotate; in other words, each pole becomes a north magnetic pole as its ac voltage reaches its peak. Therefore, there is a rotating north magnetic pole.

There is also an accompanying south magnet pole opposite to the north pole. The overall result is a rotating field. Visualize the north pole moving from one pole to the next in sequence as

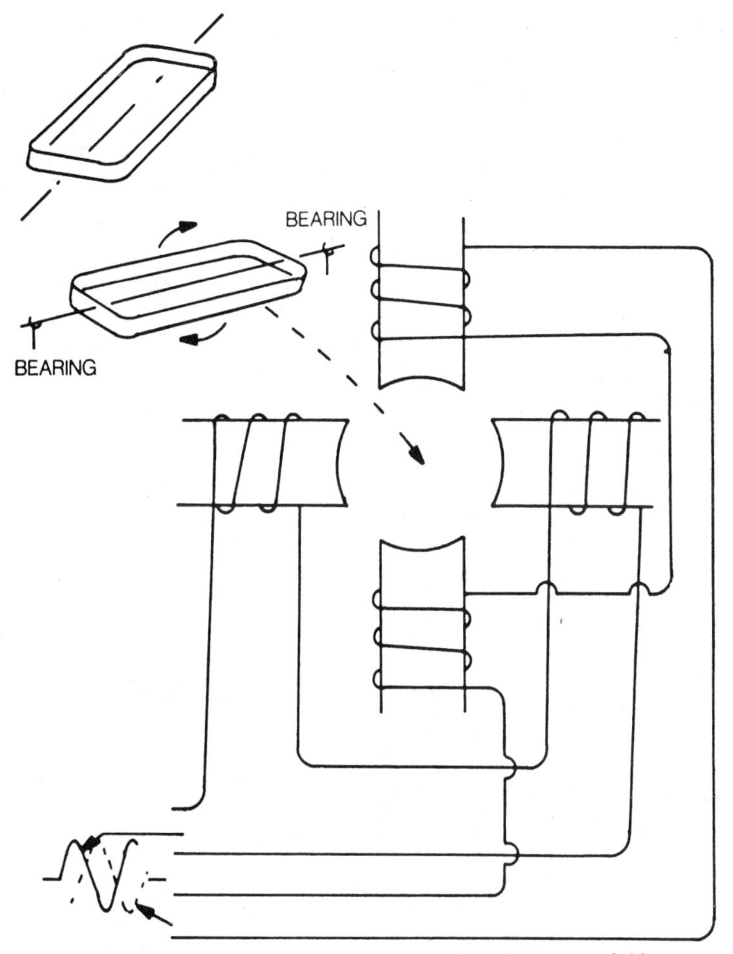

Fig. 7-21. Two-phase current can be used to produce the rotating field.

the two waveforms reach their peak. Then, corresponding south poles appear on the opposite poles.

When you place a piece of soft iron in the center of this simplified motor, it will have an induced magnetic field. That, in turn, causes it to rotate with the magnetic field.

Instead of a piece of soft iron, a conducting loop can be placed in the rotating field. It will turn if free to do so. The reason is that the moving field induces a current in the loop.

The magnetic field of the current follows the rotating magnetic field. As the loop turns, it rotates the motor shaft. In practice a number of closed-loop conductors are used for the armature.

Rotating conductors in the armature will fall slightly behind the rotating field. That is necessary because if they were in exact synchronization with the rotating magnetic field there would be no relative motion between the field and the armature. Consequently, there would be no induced current flow. When the armature follows behind in an ac induction motor, it is called slippage.

Figure 7-22 shows a more efficient set of conductors for the armature of the motor in Fig. 7-21. This illustration shows straight conductors between end conductors. They are arranged in a circular pattern. This type of armature is called a "squirrel cage." It is used in squirrel cage induction motors.

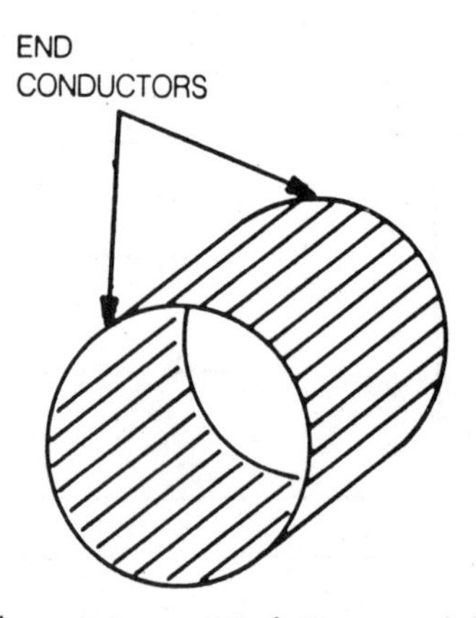

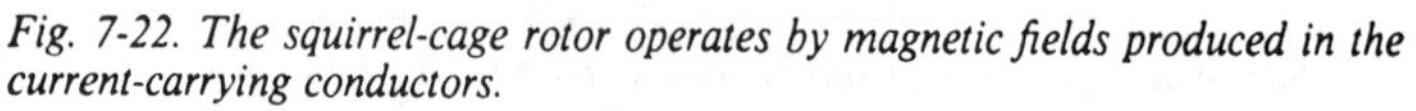

Fig. 7-22. The squirrel-cage rotor operates by magnetic fields produced in the current-carrying conductors.

The Capacitor Start Motor. So far we have reviewed two ways of obtaining a rotating magnetic field. One is by rotating a permanent magnet and the other is by using two-phase ac power. Two-phase ac power is not readily available today, and a rotating magnet does not make a highly efficient motor.

An ac motor that will run on single phase is illustrated in Fig. 7-23. It is called a capacitor start motor. Instead of delivering two individual phases to the windings, one of the windings is connected in series with a capacitor. That capacitor shifts the

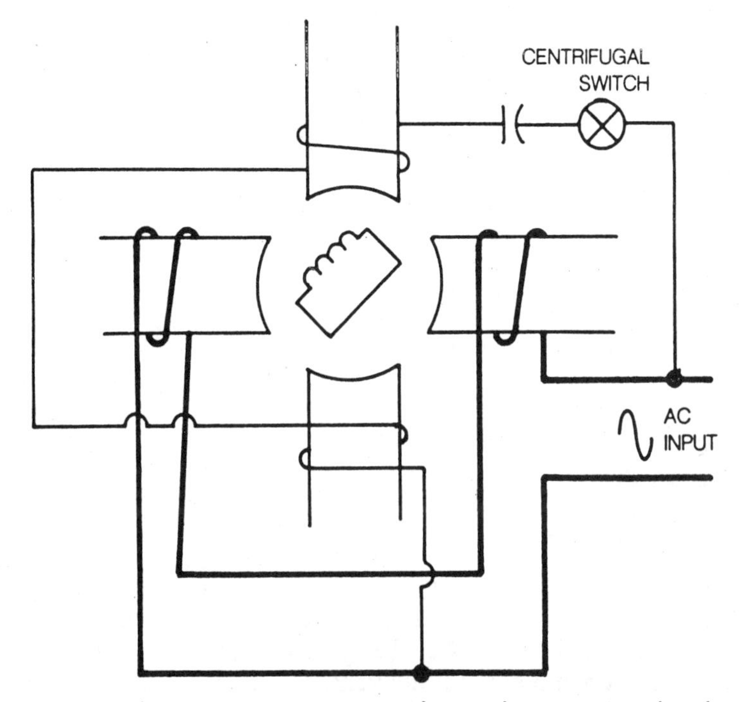

Fig. 7-23. The capacitor start motor provides two-phase operation when the motor is first started.

phase of the ac current in the capacitor start windings and accomplishes essentially the same thing as the two-phase motor.

In practice, the capacitor is only needed for starting the armature. Once it comes up to speed it is only necessary to use a single phase, synchronized so that it produces a thrust at exactly the right moment. For that reason, the capacitor start motor has a centrifugal switch. When the motor comes up to speed the switch is automatically opened and the capacitor and capacitor-start windings are removed from the circuit.

Another way of getting a rotating field is illustrated in Fig. 7-24. In this case, the field pole is split into two sections. A copper ring surrounds one of the pole pieces. Note that the pole is laminated. This is characteristic of ac field poles. It reduces the problem of eddy currents.

An ac current is induced in the single-turn coil of the shaded pole by the increasing and decreasing field of the larger

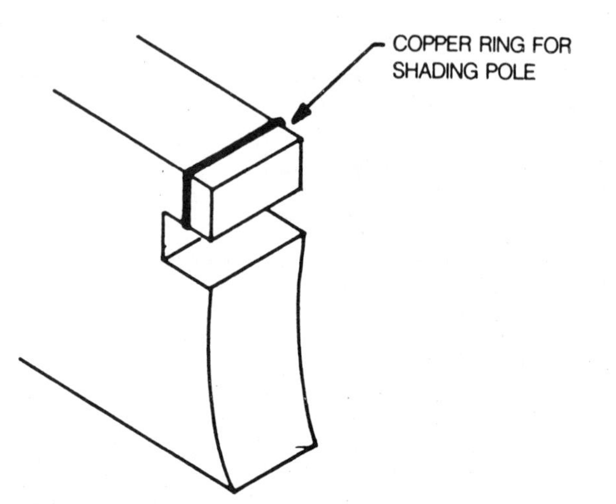

Fig. 7-24. In some ac motors, shading poles are used to produce a rotating magnetic field.

pole. The induced current is out of phase with the inducing magnetic field. Therefore, its magnetic field is out of phase with the field of the main pole. The overall result is that there is a sufficient amount of phase difference to produce rotation in the armature.

Three-Phase AC Motors and Generators. Instead of using a single pair of poles as shown in Fig. 7-21, it is possible to use three individual poles. Each pole is energized by one of the phases of three-phase power. So, there is a rotating magnetic field. Three phase ac is very popular in industrial systems. If a rotating magnet (or electromagnet) is rotated inside the three coils, a three-phase voltage is generated.

Figure 7-25 shows a basic three-phase generator and a basic three-phase motor. Permanent magnets are used in these designs, but electromagnets are used in larger units.

The Synchronous Motor. When a permanent magnet is used for the armature of a motor as in Fig. 7-26, it is not necessary to have a slippage. Therefore, once started, the armature keeps pace with the magnetic field. To get the armature up to the speed of the magnetic field, however, a modification of the squirrel-cage motor is used. This arrangement is called a synchronous motor.

Synchronous motors are used in electric clocks and other systems where the ac power frequency determines the speed of rotation. Electric clocks often use a combination of a shading

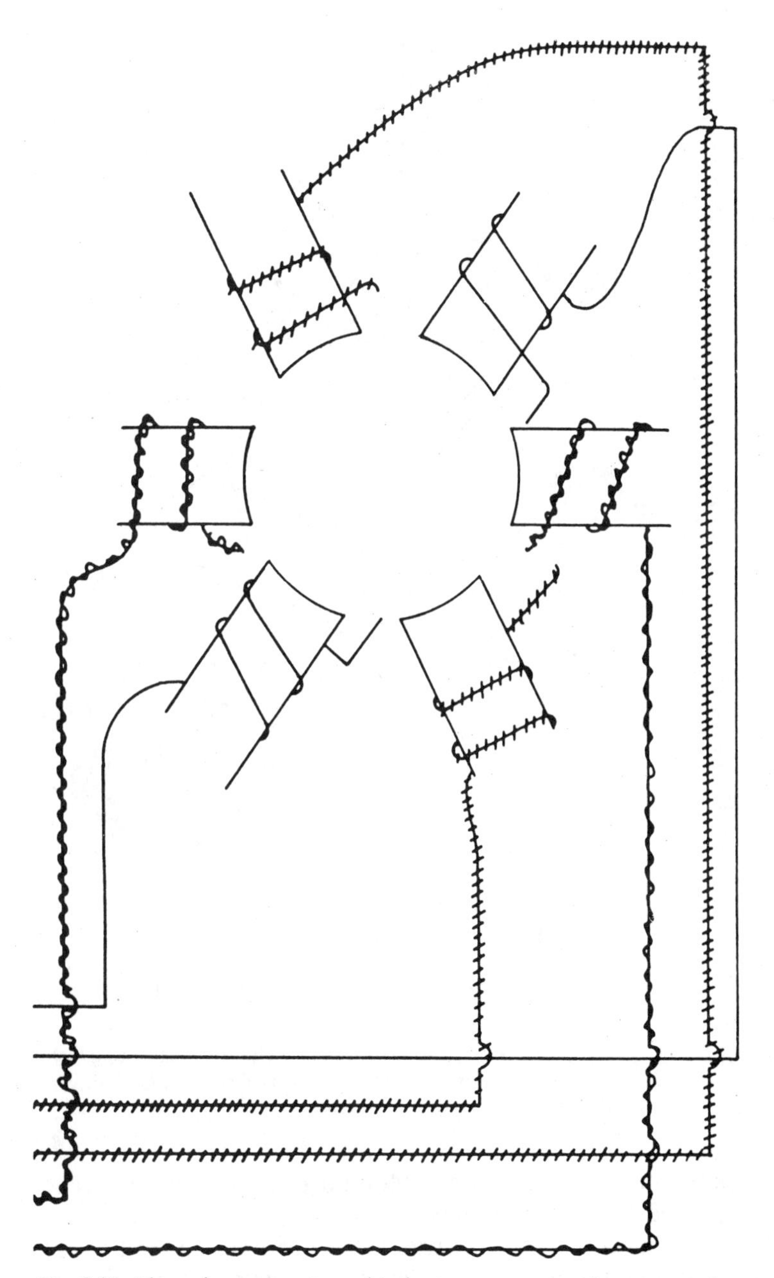

Fig. 7-25. Three-phase power is used in larger motors to produce the rotating magnetic field.

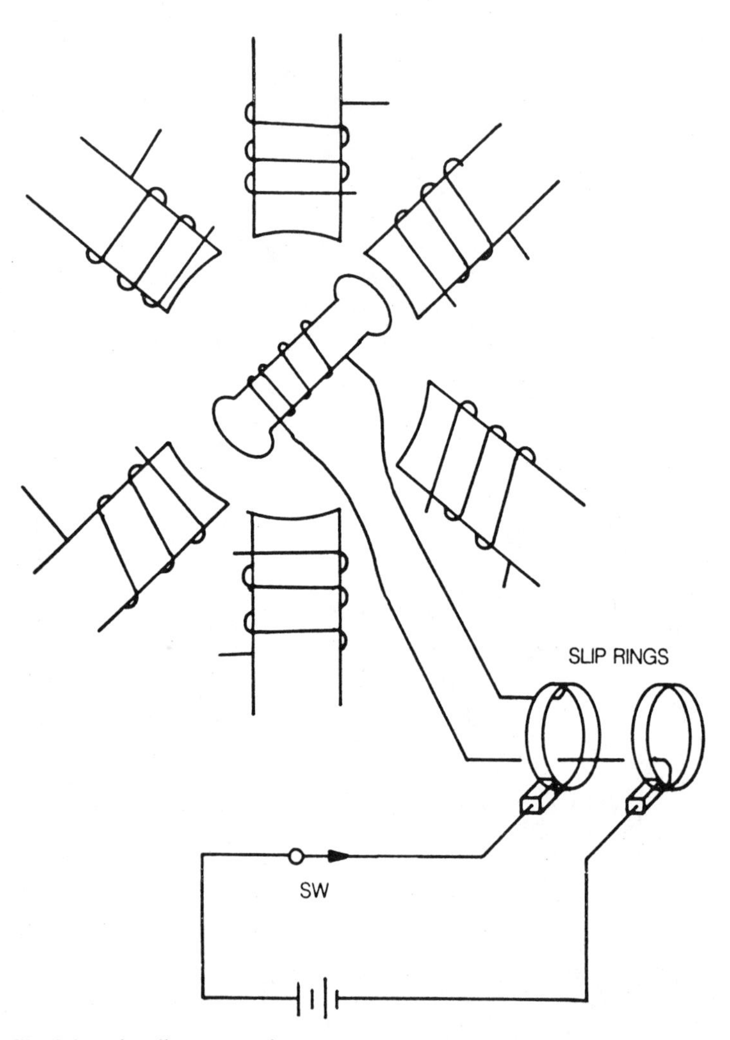

Fig. 7-26. This illustration of a synchronous motor shows that dc is supplied to the armature when the motor is up to full speed. Very often a smaller dc motor is used to get full speed in the large synchronous motor.

pole and a rotating permanent magnet in a specially constructed synchronous motor.

In a early days of electric clocks, there was no shading pole and the motor was brought up to speed with a manual thumb wheel. That produced a problem not characteristic of today's clocks. The clock could be started backward or forward depending on how you moved the thumb wheel. If the clock did not have a sweep second hand you might not notice. You could be late for work if you started the clock improperly after the ac power was off as a result of a thunderstorm. The shading pole, then, provides the correct automatic starting direction for the electric clocks made today.

AC/DC Devices. Some types of devices will work on either ac or dc. They are usually made with laminated iron material for reducing eddy currents in ac applications. But the laminations do not interfere with dc.

Consider the actuator in Fig. 7-27. It has a coil wrapped around a hollow nonmagnetic material. Inside the nonmagnetic material there is a spring and a chamber for the plunger. The plunger is made of soft iron with a steel ring attached.

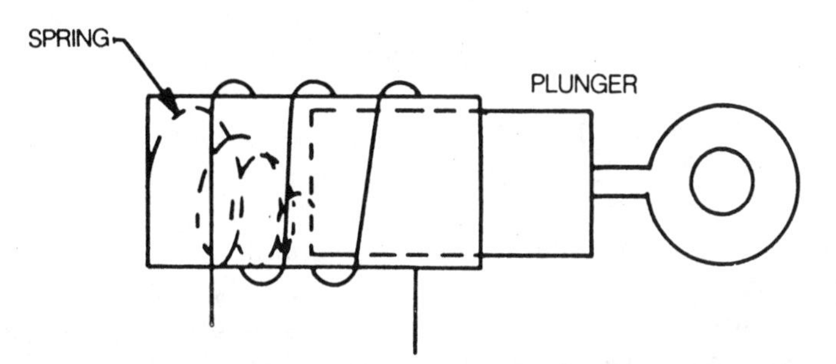

Fig. 7-27. An actuator can be made to operate on either dc or ac current input.

If dc is applied to the coil, a magnetic field will be produced. That magnetic field will attract the soft iron plunger and the plunger will move into the chamber, compressing the spring. If the ring is connected to some device that is to be moved, this action will complete the cycle of operation. When the current is interrupted, the spring will return the plunger and the ring to its zero position.

Now suppose the current is reversed in the coil. There will be a magnetic field again, and the actuator will go through the same cycle as described before, because the soft iron plunger will be attracted to a north magnetic pole or to a south magnetic pole. It does not make any difference.

Suppose now a low frequency ac current flows through the

coil. On one half-cycle the plunger is pulled in by a north pole. On the next half-cycle it is pulled in by a south pole. The current reverses back and forth with the applied ac and both half-cycles attract the plunger. There are relays and reed switches that will operate on either ac or dc. Their operation is based on this same principle.

SUMMARY

Both ac and dc motors and generators are used in industrial systems. AC motors and generators are usually made with simpler construction. Also, ac power is readily available for ac motors. AC generators can be made without the need for carbon brushes. The speed and torque of ac motors is not as easily controlled as they are for dc motors.

DC motors can be made with specific characteristics. They are especially important in portable applications where batteries power the motors. DC motors can be made to have a very high startup torque.

Both ac and dc motors and generators operate on basic laws and effects. Examples are: Faraday's law, Lenz' law, electromagnetic induction, and magnetics. Also, Ampere's law and the right hand/left hand rules are important.

Some interesting applications make use of the fact that a dc motor will generate a dc voltage when its shaft is turned. Also, a dc generator will run like a motor if dc is applied to its terminals.

Using dc motors as generators, and dc generators as motors has an important limitation. The neutral axis is different for the two devices. For efficient operation, some means is necessary to shift that neutral axis or minimize its effect.

Interpoles are sometimes used to straighten out the flux lines in dc motors and generators. This reduces the problem of the neutral axis.

AC motors have laminated cores for their fields. This reduces the problem of eddy currents. Both ac and dc motors and generators produce a countervoltage that limits armature current, and in some cases, field currents.

When large dc motors are first started, there is no countervoltage. Therefore, some method is needed to limit current in the armature until the motor is up to speed. A manual starter —or its electronic equivalent—is needed for that purpose. Even though your interest is primarily in electronics, you should understand the operation of motors and generators.

SELF TEST

1. Which type of motor will self destruct if operated without a mechanical load?
   (A) synchronous.
   (B) series-wound dc.
2. Countervoltage is the result of:
   (A) Lenz' law.
   (B) the right-hand rule.
3. One method of minimizing the effects of the neutral plane is to use:
   (A) consequent poles.
   (B) interpoles.
4. Manual starters for dc motors provide:
   (A) maximum armature current during startup.
   (B) minimum armature current during startup.
5. An alternator, like those used in cars, has:
   (A) an ac output.
   (B) a rectified dc output.
6. When the field current of a dc generator is produced by the generator it is called a:
   (A) self-excited generator.
   (B) self-biased generator.
7. In a capacitor-start motor, the capacitor:
   (A) is used to provide two-phase starting power.
   (B) is switched into the circuit when the motor has reached full speed.
8. Which of the following is used as an armature in some types of ac motors?
   (A) buck winding.
   (B) squirrel cage.

9. Which type of motor must operate with slippage?
   (A) induction motor.
   (B) synchronous motor.

10. You can get a rotating field in an ac motor by using:
    (A) a capacitor start arrangement.
    (B) three-phase power delivered to three sets of windings.
    (C) Both choices are correct.
    (D) Neither choice is correct.

ANSWERS TO SELF TEST

1. (B)
2. (A)
3. (B)
4. (B)
5. (B)
6. (A)
7. (A)
8. (B)
9. (A)
10. (C)

8

Robotics

CONSIDERING ONLY ELECTRONIC circuitry, there is very little difference between robotics and any other control system. For example, electronic components used for the circuits in robots are the same as those reviewed in the first few chapters of this book. Among the important parts of a robot system are transducers, motor speed controls, and power supplies. They are identical to those that have been discussed earlier.

The first readily available robot for industry was marketed in 1959 by a company called the Planet Corporation. From that time until today the electronic technology used in robotics has simply kept pace with the development of electronic technology in general.

Why then is a chapter on robotics important? Because over and above the electronics, there is some terminology that you should understand. As a robotics technician you would be expected to be able to read the company manuals and literature in publications. In other words you would be expected to keep abreast of the technology as it applies to robotics. You can only do this if you understand the terminology used in robotic systems.

Some important advances that have contributed to the advancement of robotics have come from the aerospace program and other progressive systems that employ electronics. From the time of the first robot, electronics has progressed through the vacuum tube era to modern microprocessor and digital electronic systems. The history of robotics technology is simply a history of electronic technology. But, there have also been tremendous improvements in such areas as metal alloys and computer control.

Unfortunately, robots have caught the imagination of science fiction writers and writers for the media. They have been projected into uses for which they are totally unsuited today. Yet the possibility exists that some of those things can be done in the future. One of the most exciting of these is the anthropomorphic robot that closely simulates human motions.

Figure 8-1 shows the types of motion that a human arm can easily perform. Robots are often identified by which lettered axes the arm (or moveable part) can move through.

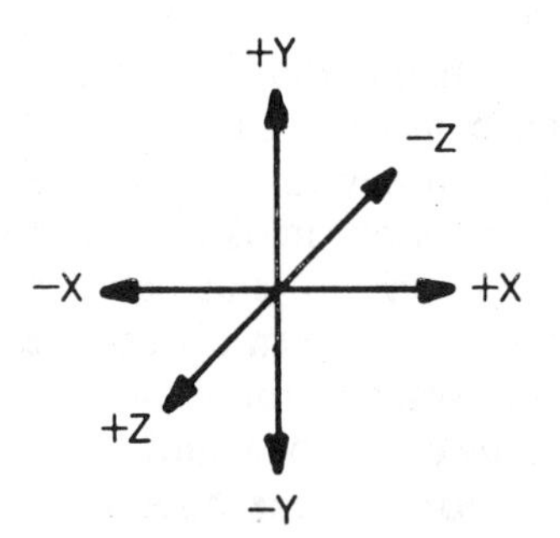

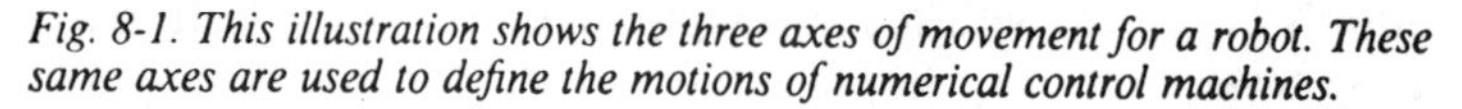
Fig. 8-1. This illustration shows the three axes of movement for a robot. These same axes are used to define the motions of numerical control machines.

In real life, robots can be very simple, and they may in fact contain no electronic control at all. These may not be as exciting to read about, but they are every bit as important in industrial applications.

Study the terminology carefully in this chapter. That, after all, is the most distinguishing feature of robotics compared to other electronic systems.

CHAPTER OBJECTIVES

After you have studied this chapter you will understand

- The LERT classification of a robot.
- What determines the work envelope of a robot.
- The bang-bang robot.
- Some other terms that describe the robotic manipulator.
- Why pneumatic and hydraulic robots are preferred over electronic robots in some applications.

ROBOTICS: THEORY AND PRACTICE

A robot may be designed to move a tool such as a drill, a welder, a paint sprayer, or a part. Regardless of what it is designed to move, the operation of the robot *arm*—that is, the moveable part of the robot—is essentially the same. Also, the method of programming such a robot can be the same for different jobs.

Programming a robot depends somewhat upon the complexity of the work it is to perform. It is much simpler to design a robot to pick a part out of a basket and deliver it to a point, than it is to design a robot to follow a complex contour that has to be welded.

Programming—The Key to Flexible Operation. The word *program* is not unique to electronics. A program is simply a step-by-step procedure. For robots, computers and other electronic systems, that procedure is usually in the form of a binary code stored in some kind of memory device. In addition to computer memories, storage can be on magnetic tapes, disks, punched tape, or some other method of holding the program until it is needed. Computers microprocessors, and some digital systems are programmed.

Non-electronic systems can also be programmed. A simple way of programming a robot movement is to set up a stop that it moves against. The stop is a mechanical obstacle that simply prevents the movement from continuing.

If there is one word that distinguishes robots from other devices it is versatility. A robot must be flexible in its operation so that it can be assigned different tasks within its capability. Its capability, of course, includes how heavy an object it can lift.

Another important determination of its capability is its range and type of motion. For a specific location in industry, a robot must be able to handle a wide variety of jobs.

Take, for example, the case of welding a non-uniform complex contour in an application where one hundred welds are necessary. If this is a job of welding in the automotive industry, the yearly change of models would put the robot out of work unless it could be reprogrammed to handle the new welding contours.

Trying a Specialized Application. Despite the excitement created by anthropomorphic robotic systems they cannot compete with a human in some of the most simple tasks. Consider this very simple task for a human being—picking an egg out of a basket of eggs, turning it to the right position, and putting it in an egg crate with the largest part of the egg down to fit an egg crate opening. Even a child can be taught to do this, although it may take a few tries. But, this is a very difficult task for a robot.

This is not to say that a robot could not do it. However, the cost and complexity of the robot would far exceed the importance of the task. So you can see it is not true that robots have released human beings from simple repetitive, and often boring tasks.

Make the job even more complicated. In a basket there are golf balls and eggs. A human being would have no problem distinguishing between the two and you would not end up with golf balls in the egg crate. Making the same distinction in robot operation is an enormous task and involves an enormous amount of memory, so that the robot would reject not only golf balls, but other objects that are similarly shaped and colored.

If you gave the job of loading the egg crates to a human, the human would stop when the egg crates are empty. A robot, however, knowing only the job of picking things out of the basket and putting them into the crate, would continue to work into the night even though the basket was empty.

Of course, using special electronic techniques, the robot could be trained to look for eggs, and on not finding any, to shut itself off. But then, you would have to make provisions for it to start up again when a new box of eggs is placed in position. That would involve additional memory circuitry.

If you used a robot for this task, the cost of eggs would be prohibitive. The cost of the robot would have to be spread over years. Accountants like to call this *amortizing* the cost. By the time it was nearly finished paying for itself, it would be time to look for a newer, better, faster, cheaper robot, so that maintenance would not be such an important factor added to the cost of the eggs.

Some Advantages of Robots. From the example you might wonder if it is ever necessary to have a robot. However, there are applications where robots must be used instead of humans.

- A robot can go into a room full of poisonous gases and shut off a valve. By the time a human gets suited up for a job like that, the poisonous gases could have done an enormous amount of damage.
- A robot can work in highly dangerous areas such as places where explosive devices are assembled. If an explosion took place, a robot could be replaced, but not a human.
- A robot can do a boring, repetitive job day after day after day, without getting discouraged or developing some psychological problem that has ramifications in society. Human beings usually rebel at that kind of work. They just simply do not want to do it. It is better to use a human for other types of work.

TYPES OF ROBOTS

Human beings have some basic characteristics in common, yet every human being is different from all others in some ways.

You can say the same thing about robots. They all have in common the types of work they perform and other basic features, but there are some distinguishing characteristics used to identify specific kinds of robots. These are classifications. They will be listed here and then discussed in greater detail.

One classification is called LERT. The letters stand for Linear-Extentional-Rotational-Twisting. These are the four basic types of robotic motion. LERT robots can handle all four of them.

Another way to classify robots is by the geometry of their

arm motion. In the simple robot of Fig. 8-2, it can only move in two directions. The end of the arm will trace out a rectangle at its extremes, so it is called a rectilinear, or rectangular robot. A third way of classifying the robot is by the amount and type of technology involved in its operation.

The LERT System. The LERT system defines a robot in terms of its ability to move within a work envelope. The envelope is the total extension of the robot in all directions. The envelope can be rectangular, as in Fig. 8-2, but it can also be cylindrical or spherical. The rectangular pattern is viewed from the top. The LERT classification defines the directions it can move, such as linear, extensional, rotational or twisting. It also defines the type of envelope obtained by those motions. Motion is always defined starting at the base and working toward the outer limits of the robot extension.

Rotational. Figure 8-3 shows a robot with rotational mo-

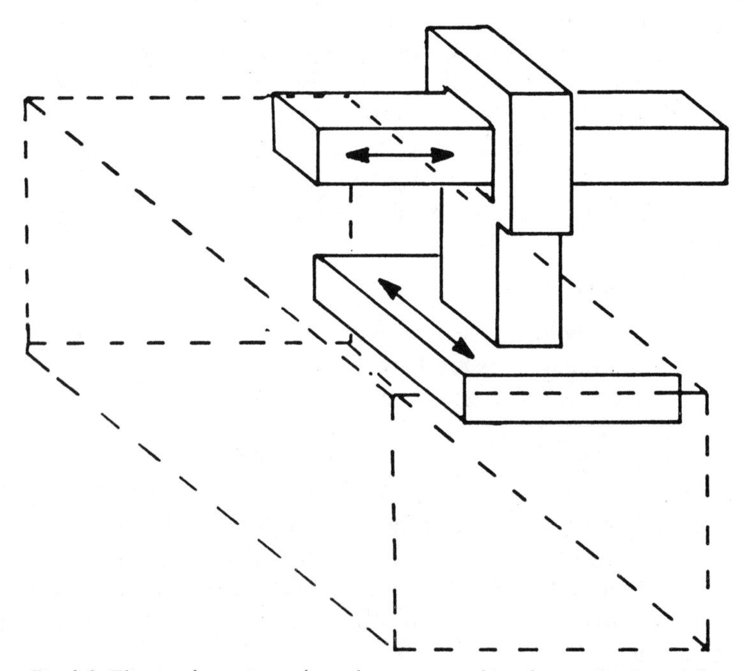

Fig. 8-2. The simple motions, shown by arrows on this robot mechanism, define a rectangular envelope.

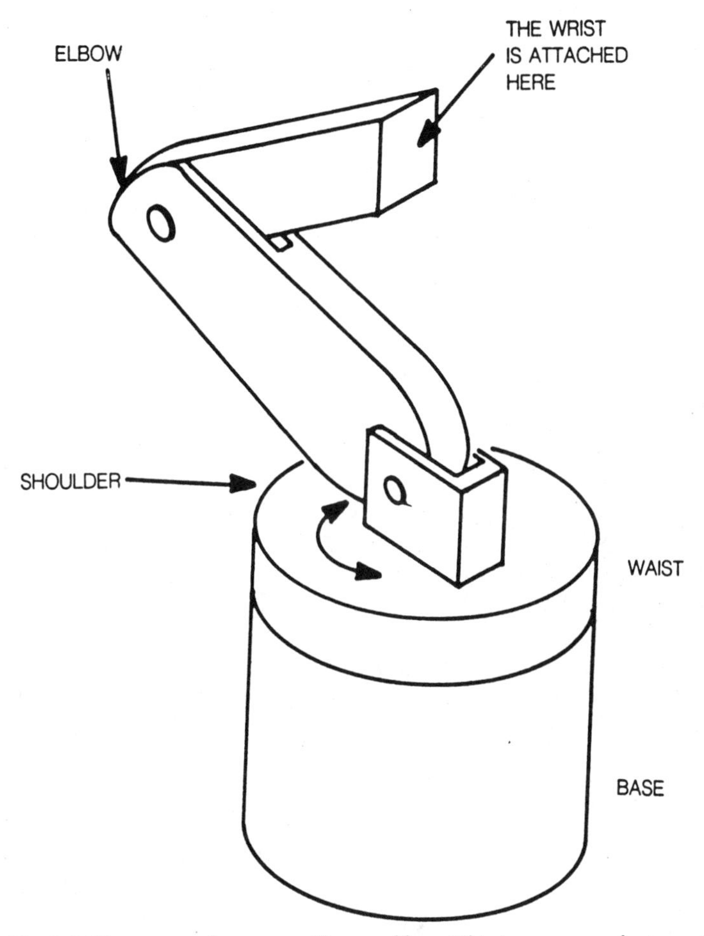

Fig. 8-3. Three types of motor are illustrated here. This is a very popular type of robotic arm.

tion at the base. The stationary part of the base is called the base and the rotational part is called the waist.

The rotational motion of a robot is shown from the top in Fig. 8-4. It is normally less than 360 degrees. The extensional motion of a robot can be seen in the side view of Fig. 8-5. The extension of a robot arm is no different from the extension of your arm. This can also be called the reach.

In robotic terminology there can be two types of rotation in the same robot. This is called rotational-rotational-extension,

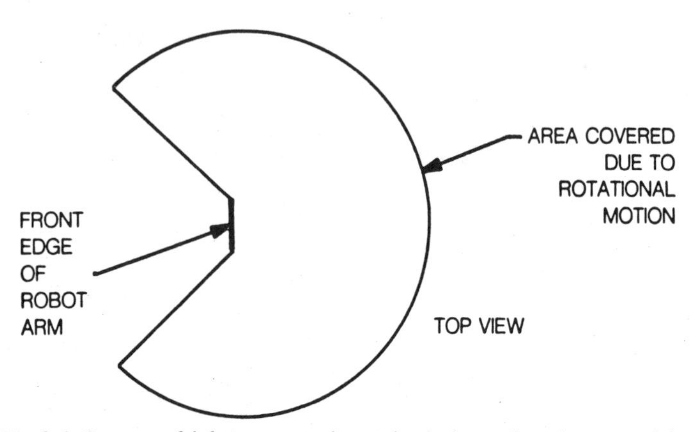

Fig. 8-4. One way of defining an envelope is by the limit of reach as viewed from the top.

and is identified by the letters RRE or R2E. The twisting motion of a robot occurs as a result of its wrist motion. This is illustrated in Fig. 8-5. To distinguish between rotational and twisting motion, remember that the twisting motion does not affect the work envelope of the robot.

Classification by Geometry of Robot Arm Motion. As described before, a robot may have a work envelope shaped like a cylinder, a rectangle, or a sphere. So, it can be classified as

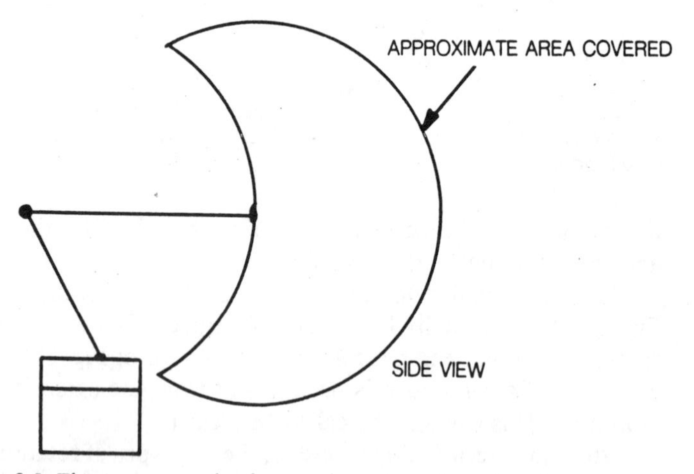

Fig. 8-5. This is an example of an envelope viewed from the side.

being a robot of the type of geometry that it can trace with the end of its manipulator (HAND). Remember that the envelope is the limit of the robot's reach at any point around it (Fig. 8-6).

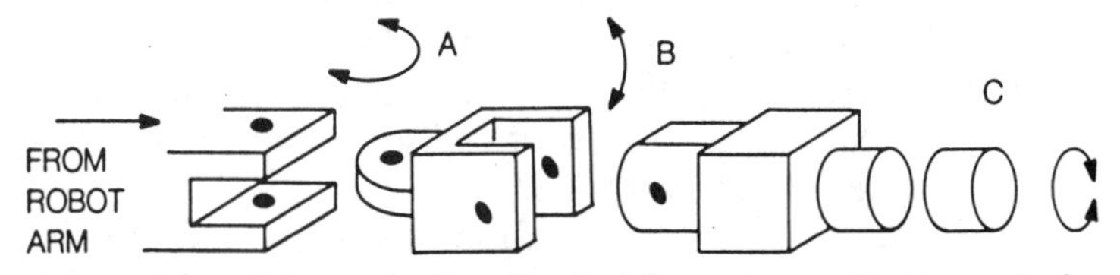

Fig. 8-6. Some of the mechanisms for obtaining various motions on a single robot arm are viewed here. A permits motion on the x-axis and B permits motions on the y-axis. C provides wrist motion.

Classification by Type of Technology. There are three ways to move a robot: electrical, pneumatic, and hydraulic. Each of these methods has its own application and range of operations.

Hydraulic Types A hydraulic robot is usually one capable of lifting very heavy payloads. Normally, hydraulics are not used for small robots.

Although they can lift the heaviest payloads, hydraulic robots present a problem in terms of continued maintenance. The hydraulic fluid must not be permitted to overheat. The fluid must be filtered to keep it clean, and it must be returned to the hydraulic pump to be recirculated. In this, hydraulics are very much like electrical systems. In electrical systems the current won't flow out of the power supply unless it has a way of getting back. Likewise, the hydraulic system provides continual maintenance to the fluid and filters in the system.

There are also periodic maintenance problems with pumps and other moving parts. However, the maintenance problems of hydraulic systems are not necessarily any greater than for electrical or pneumatic systems.

Pneumatic Types Pneumatic systems use air to achieve the motion of the robot system.

Pneumatic systems have a very important advantage. The air can be vented rather than having to be returned to the source. That makes them simpler to design and cheaper to manufacture. However, compared to a hydraulic system the use of air has an important disadvantage. Hydraulic fluids are noncompressible but air is greatly compressible. This affects the maximum payload that can be lifted and the accuracy of positioning. Increasing the payload decreases the accuracy. As a rule, maintenance problems are less, but at the same time the maximum payload is not as great as with hydraulic systems.

Electrical and Electronic Robotic Systems In terms of accuracy, electrical and electronic robotic systems are superior to other systems. By using stepper and servo motors, the robot can be very accurately positioned.

Accuracy is great because of the extensive electronic controls that can be applied. Accuracy provides repeatability, which is absolutely necessary for some types of operations.

Servo control is another name for feedback control. It is one of the features responsible for the great accuracy of electrical and electronic systems. Feedback can be used to provide the system an exact position and speed of the robot manipulator. The manipulator (hand) is used for grasping an object. In its simplest form, it has two parallel surfaces as shown in Fig. 8-7.

On the negative side, electrically operated robots cannot handle the heavier payloads of pneumatic and hydraulic systems. If you are to list them in a descending order of payload, it would be hydraulic, pneumatic and electrical. Of course, that is not always a great disadvantage. Just as you wouldn't use a heavy tractor and trailer to deliver a diamond ring, it would be unnecessary to use large, powerful robots for such delicate operations as placing an electronic component in a printed circuit board.

Electronic sensing components such as thermistors and light activated components can serve as the sense of touch and sense of sight for the robot if the robot is used for picking eggs out of a basket and placing them into an egg carton. This job would have to be done with a robot that employs electrical power and an electronic control system.

Degrees of Freedom for Wrist Motions. If you are classifying a robot in accordance with its type of motion for its envelope, it is important to list its *degrees of freedom*. This is one way of defining the axes of motion for a robot. There are two categories for degrees of freedom. The major axis category involves side to side motion, up and down motion, and reach. Basically, this specification is a list of degrees of motion or freedom that defines the envelope of the robot. The minor axis

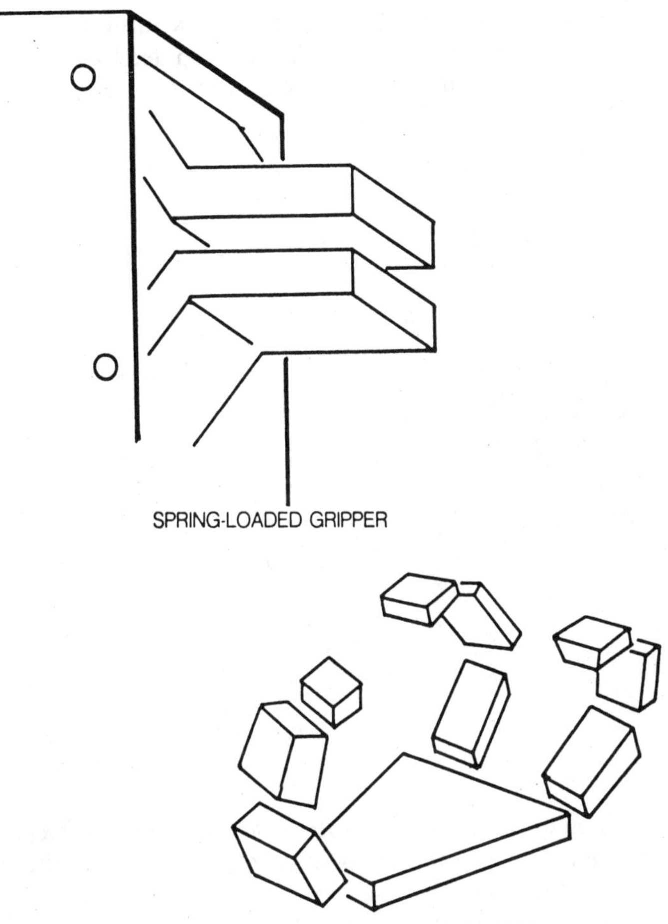

Fig. 8-7. There are two types of end effectors illustrated here.The spring loaded gripper is the simplest, but is very limited in its capability. The anthropomorphic gripper can simulate motions of a human hand.

for degrees of freedom is determined by the robot's wrist motion. It includes roll, pitch, and yaw.

You can easily visualize these motions by holding a ruler in your hand at arms length. Remember that the wrist motion of a robot has no effect on the envelope. So, when you place the ruler at arms length, realistically you are operating the ruler in the minor degrees of freedom.

Hold the ruler so that it is vertical. By turning your wrist you can rotate the ruler clockwise and counterclockwise. Here you are exerting the roll function, illustrated in Fig. 8-8.

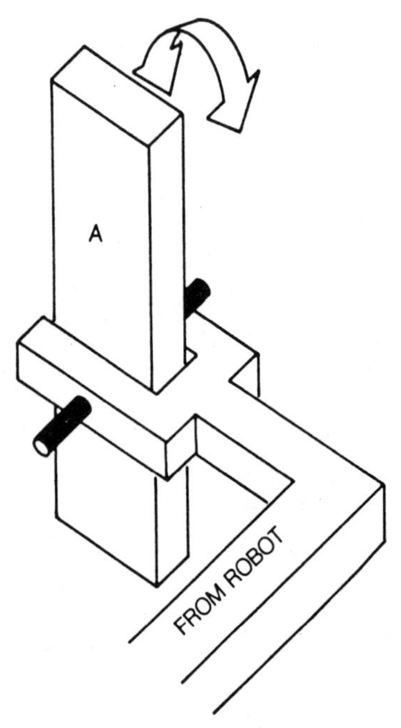

Fig. 8-8. This is another way to obtain wrist motion in a robotic arm.

You can move the ruler from side to side while keeping it in a vertical position. This is an example of a yaw of the wrist motion. (See Fig. 8-9).

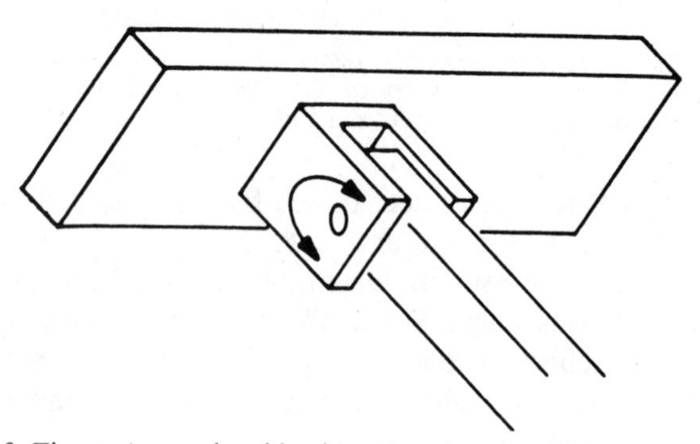

Fig. 8-9. The motion produced by this wrist action is called yaw.

Finally, you can move your wrist so that the bottom of the ruler points toward you or away from you (of course the top will go in the opposite direction). This is an example of the pitch motion. (See Fig. 8-10.) Not all robots have a wrist motion that permit them to move in the minor axes.

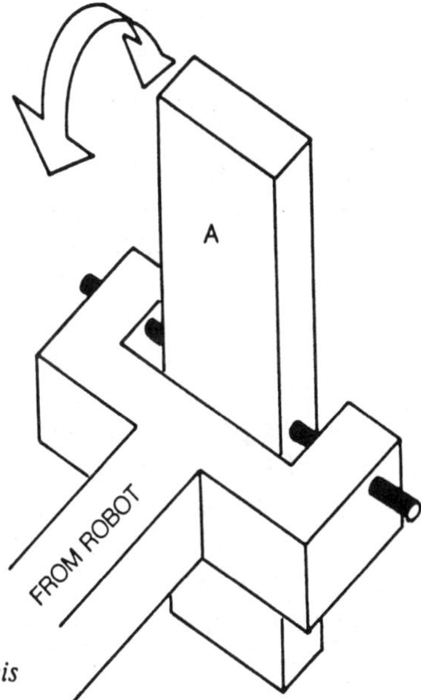

Fig. 8-10. The motion produced by this wrist action is called pitch.

DIVISIONS OF ROBOTIC TECHNOLOGY

If you were going to design a robot, there are three major areas that you would have to consider.

First, you would have to consider the power supply. That doesn't mean the same thing in robotics as it does in electronics. In electronics the power supply delivers the power necessary for operating the electronic components. But in a robotic system the power delivered to a robot can be either electric, pneumatic or hydraulic.

The second area to consider would be the manipulator. The manipulator, or hand, of a robot is also called the end defector.

Finally, you would need to have some way of controlling the robot. The controller is the third area of concentration, or area of specialty you would have to consider when designing a robot.

The Power Supply. The type of power supply is dependent greatly upon the kind of operation you are performing. Certainly, if you are going to have a hydraulic system you would not employ elaborate controls when that system is being used for very heavy payloads.

Consider the example of the robot that is moving cars onto a conveyor at a junkyard. The robot drops down to the roof of the car and an electromagnet lifts it. It does not matter if the electromagnet is an inch right or left, or front or back of the car. That degree of accuracy is not necessary.

If you are going to insert an electronic component into a printed circuit board, you would need some very precise electronic controls and some feedback circuitry to manipulate the part into the proper position. Consider the size of the holes and the size of the leads of the component. Mentally imagine that you are placing these parts into the board. If you are going to do it quickly you will need great accuracy in aiming.

Hydraulic and pneumatic systems are sometimes employed in a type of control referred to as "pick and place," or "bang-bang robots." They get that name from the fact that they are limited in their travel by banging against mechanical stops. Instead of feedback, the mechanical stops determine the maximum travel of the robot arm. This type is noisy but it is highly reliable. Also, since electronic or elaborate feedback circuitry is not needed, it is very inexpensive, and it is easy for a relatively untrained technician to set up.

Point to Point. Servo controls can be either electric, pneumatic or hydraulic. There are three distinct types of servo control.

A point-to-point servo control permits the robot to be moved from one specific point in its envelope to another specific point. This is a more complicated procedure than the bang-bang method.

Return again to the problem of picking the egg out of the basket and putting it into an egg crate. The arm will have to turn and reach down into the box. Two axes of motion or two degrees of freedom are necessary there. Then, in lifting it, it must be able to not only turn and raise the egg, but it must also be able to extend the egg to position it over the egg crate. Furthermore, it must be able to turn the egg so the largest part of it is on the bottom.

If you were to trace the path of the end effector as it goes through this motion, you would see that it can be quite complex in a servo system. The control might be able to move the robot in two or more directions at the same time.

The point-to-point robot is the most popular type used in automatic fabrication. It is one of the most "intelligent" robots. In order to get the robot to perform this complicated task, it must be programmed. Sit in a chair and move your hand as though you were picking an egg out of a basket on the floor and placing it in an egg crate on the table.

Additional Notes on Programming. One method of programming is to move the robot through its required motion using a teach pendant. A teach pendant is nothing more than a remote control. When the operator moves the robot with the teach pendant, each motion is stored in memory. Assuming the operator is satisfied that the robot is moving through the necessary positions, a switch on the teach pendant called *run* is pushed. The robot can then go through the same motions without the need for a human to direct it.

Teach pendants are similar in some ways to remote controls for television sets. For example, you set the remote control for a certain amount of sound volume. When the set is turned off and turned on again, it "remembers" that same volume position, where operation will start again. You can reprogram it, but it always remembers what you programmed for its starting position. If you select channel six with a remote control and turn the receiver off, it starts at channel six when it comes on again.

The teach pendant does a similar job. It tells the robot where to start and when to finish. In the hands of a capable programmer, the robot will go through its motions efficiently.

Since the shortest distance between two points is a straight line, the person operating the teach pendant will move the part most quickly by going along a straight line. Of course, the path that it follows must be unobstructed.

Another method of programming a robot is to simply hold its end effector and move it in exactly the direction it is expected to go by itself. As the end effector is moved in this lead *through programming*, the motions are stored in memory. Actually, it does the same thing as the reach pendant. It remembers the motion by using large amounts of memory.

Both of these methods of programming are applications of point-to-point. They require that the assembly line, or other location of the robot, be called to a halt while the programming takes place. This can be very expensive. The amount of down time is costly because the product is not being moved, and a finished product is not possible. Also, assembly line workers will be paid even though they are idle.

A method of programming that does not require that the assembly line be shut down is called on-line programming. This method uses a computer in a remote location. A computer aided design (CAD) computer program is adapted for this job. The operator can completely program the system while observing the motions on a computer display screen. The computer puts all the required motions into program memory. Then, this memory content can be dumped, or down-loaded to the robot system with a minimum amount of changeover time.

It costs more money for this kind of programming, but the overall cost, which takes into account the loss of production time, can be greatly reduced. For that reason this type of programming has become very popular.

Continuous Path Motion. With continuous path motion, the robot simply goes from one position to the other against mechanical stops. You will recognize this as the bang-bang method. It is the cheapest and easiest to make. Do not overlook the effect of the noise on personnel working near this type of robot.

The control path method is very similar to the point-to-point, but it does not have to be quite so accurate. You would use a control path for moving a paint sprayer. The programming can be exactly the same as previously described.

A feature of the continuous path robot is the fact that it can be made to move in a straight line between two points. This makes it possible to get the job done more quickly. It is different from the control path, in which the robot follows its own natural path in moving something from one point to another.

If you pick up an electronic component and place it on a printed circuit board, you are not normally conscious of the direction and path that the part takes in going from the box to the printed circuit board. You simply pick it up and place it there. Your arm follows a natural motion.

To move the part most quickly, you would pick a straight line path between the box and the device. To summarize, the continuous path is not the fastest, but it is natural. The control path gets you there quickest.

END EFFECTORS

End effectors, or hands, are also called *grippers*. They are the most complicated of all tooling on a robot. They are placed at the end of the robot arm. Grippers come in a wide variety of configurations, but in general they all grip, lift, and then release. Grippers like the one in Fig. 8-7 can grasp from the outside or inside of a hollow part. These actions are controlled by the robot controller.

Some important technologies are often involved in making end effectors. For example, they may employ infrared devices that make it possible to sense the location of an object. Also, they may be able to perform in any or all of the positions in the minor degrees of freedom.

The simplest grippers are spring operated. In that case, they are either held closed or held open with a spring until a signal is delivered. This type cannot be used for sensing. In other words, you wouldn't want to use one for lifting an egg, because there is no feedback into the system to indicate that touch has been performed.

Other grippers employ infrared and ultrasonic technology which sends signals back into the system making a much more delicate operation possible.

One type of gripper employs a suction or vacuum to lift the work piece. These are capable of handling very delicate parts and objects but it is not possible to regulate the force, as it is in the case of transducer sensors. These pneumatic systems are either on or off, go or no go.

You will remember the electromagnet used to lift cars in a junkyard. That is another form of gripper. When it is in place, a high current is circulated through the electromagnet that produces a very strong magnetic field. This type of end effector is very useful for heavy magnetic materials. If the material is not magnetic then it is sometimes possible to use an electrostatic device that performs the same purpose. Instead of a magnetic field, and electric field is used to lift the payload.

The most versatile of all end effectors are interchangeable. In one application, a polisher may be used. In another, a drill or a paint gun. Grinders and other industrial tools are also possible. The end effector can be changed manually by technicians on the work floor at the location, or it may be changed by the robot controller himself.

A FINAL NOTE ON SAFETY

Robots are dumb, unthinking, insensitive devices. But, they can operate with great amounts of power. For that reason, extensive safety procedures must be used to prevent the robot from hurting a human being. If you are going to work around robots, you should always keep in mind that like any machine, they are capable of seriously harming humans. Protect yourself and follow the rules and guidelines set by the company to protect you.

SUMMARY

Commercially available robots, as we know them today, were first available in 1959. Since that time the development of robots has paralleled the history of technology. New materials, new computer systems and memories, and improved control systems have all contributed to the improvement of robots.

Robots range in complexity from the simplest bang-bang type that can only move between stops to anthropomorphic robots that very closely simulate human activity. As the robot increases in complexity, so does its cost. So the type of robot used for a particular application is largely dictated by the type of robot needed. It is also influenced by the robot's price.

There are several ways to classify robots. One method is by

the envelope that the end effector traces. This envelope can be a simple rectangular type or a complex cylindrical envelope. Classification by envelope is often done by the LERT system.

Robots are also classified by the type of control that they use. The bang-bang type uses mechanical stops to limit travel. Sophisticated electronic types use electronic control circuitry.

Robots are sometimes classified by the types of jobs that they do. One example of this classification is pick and place. It is a relatively simple type.

Robots are also classified by the kind of power they use. The three kinds of power are hydraulic, pneumatic, and electric. Each has its advantages and disadvantages. For example, hydraulic robots are capable of lifting the heaviest payloads while electric robots (which are electronically controlled) are the most accurate at placing the end effector.

If you believe the stories told in the media, you may come to the conclusion that robots can replace humans in all tasks. Surprisingly, they cannot replace humans in some of simplest tasks. In fact, a human can out-work and out-think a robot—at least for a limited period of time.

The example used in this chapter was that of taking eggs out of a basket and placing them into an egg crate. This cannot be done by a simple robot. In the first place, the robot would not know to quit when the basket is empty. Secondly, if somebody put a golf ball in the basket, the robot would not know the difference between the golf ball and the egg. Furthermore, it is very difficult to get a robot to pick up an egg without crushing it. To position the egg in the egg crate so that the larger part of the egg is on the bottom is a very complicated task for a robot. The egg must be turned and there must be a special sensing system to determine when it is in the right position. A human finds very little trouble performing this task. It is doubtful if the cost of the robot would be justified for such an application.

Yet there are places where robots have an advantage. They are able to do simple repetitive tasks for long periods of time without getting bored or tired. Assuming they are properly set up, they do not get tired and make mistakes. They do not need coffee breaks and vacations, but they do need periodic maintenance and repair. Also, they must be programmed before they can do their job.

Programming is a very important part of robotics. Three of the most important methods of programming are: 1) The teach pendant, which permits an operator to move the robot through its activity. Once this has been done, the program can be dumped onto robot memory on the assembly line.

2) The lead-through method. The technician holds the end effector and actually moves it through the paths that it is expected to take. Again, having moved it manually, the motions are memorized in electronic memories and dumped to the robot system on the assembly line.

3) The computer method of programming robots has the advantage that the assembly line does not have to be shut down for long periods of time. Programming can take place at a remote location. Then, once the program has been completed it can be down-loaded to the robot system.

There are two very important paths that a robot end effector can take. One is the continuous path, which is the type of motion that humans usually follow. The other is a point-to-point straight line path, which is unnatural for a human, but can be set up for the fastest robot operations. End effectors vary from the simplest spring loaded type to very complicated anthropomorphic types that look like human hands.

Be careful when working around robots. THEY CAN HURT YOU. You must follow the rules of safety to avoid personal injury. Remember that a robot is an unthinking piece of equipment, so you must be alert to avoid accidents.

SELF TEST

1. One thing that limits the positioning accuracy for a pneumatic robot is:
   (A) the accuracy of the valves.
   (B) the compressibility of air.

2. The twisting motion of a robot occurs in its:
   (A) elbow.
   (B) wrist.

3. Which of the following types of robot does not require a return system?
   (A) pneumatic.

(B) hydraulic.
(C) electrical.

4. Is the following statement correct? A true robot is reprogrammable.
   (A) The statement is correct.
   (B) The statement is not correct.

5. Which of the following best describes an important feature of all robots?
   (A) flexibility.
   (B) electronic control.

6. A step-by-step procedure for operating a robot or computer is called a:
   (A) teacher.
   (B) program.

7. In electronic systems, a program is usually written:
   (A) as a binary code.
   (B) in the metric system.

8. Which of the following is correct when considering a short run?
   (A) Humans can perform simple tasks better than robots.
   (B) Robots can perform simple tasks better than humans.

9. A robot that can simulate human motion is called:
   (A) anthropoligec.
   (B) anthropomorphic.

10. To justify the expense of a robot, its cost is:
    (A) amortized over a period of time.
    (B) less than the cost of a human over a period of one month.

ANSWERS TO SELF TEST

1. (B) The fluid in hydraulic robots is not compressible but the air in pneumatic robots is definitely compressible. That means that positioning the arm is not highly accurate because it is hard to repeat a movement exactly. Furthermore, the compressibility of the pneumatic air can cause bouncing and oscillation.

2. (B) You might have difficulty with this answer if you tried to reproduce the robotic action by extending your arm. However, robots can hold the elbow section of their arm steady while rotating the wrist only.

3. (A) In a pneumatic system the air is vented to the space surrounding the robot. It does not have to be returned to the pneumatic pump. Hydraulic fluid must be returned to the pump and likewise, in electrical systems, there must be a return path from the power source.

4. (A) Being reprogrammable is one of the important features of a robot.

5. (A) Not all robots have electronic controls, but they certainly must be flexible so that they can be used for many different types of jobs where they are located.

6. (B) While the term "program" is applicable to all robots, it is most directly applicable to those with electronic control.

7. (A) Computers and controllers can only recognize binary numbers written in the form of code.

8. (A) Surprisingly, humans can outperform robots in short runs. One reason is that the time taken to set up the robot program—added to the time for doing the job—is longer than the time required for a human to perform the task. This is true even though some training may also be needed for the human. As a general rule, humans can learn simple tasks quicker than robots.

9. (B) This is the definition of anthropomorphic.

10. (A) However note that this is not an electronic term. It is the term used by accountants.

9

Circuits and Circuit Theory

IF YOU ARE considering a specialization in industrial electronics, then this book has introduced you to the various facets of that subject.

Over and above what is covered by courses in general electronics, there is specialized knowledge of network theorems and laws that you should be familiar with. These subjects are covered in this chapter along with some practical applications.

Some of this material will be a review—depending upon where you learned your basic electronics. However, the information in this chapter is important enough that a review is in order.

Circuit analysis is useful for understanding the construction of circuits. It is also useful for tracing circuits as part of a troubleshooting procedure. One way to give your troubleshooting techniques a broader base is to study basic theorems and laws that are related to electronic circuits. That material is covered in this chapter. We will not go into the mathematics of circuit analysis.

Resistance-capacitance (RC) circuits and resistance-inductance (RL) circuits are very important in industrial electronics. You are no doubt aware of basic time constant circuits, but they will be reviewed very briefly here. Differentiating and integrating versions of op amps will be considered already discussed.

CHAPTER OBJECTIVES

After reviewing the material in this chapter you will better understand the answers to these questions:

- What two circuits can be used to replace any four-terminal passive network?
- For the transistor parameter h_{FE}, what does the letter h stand for?
- Can active four-terminal networks be represented by impedance parameters?
- What is the relationship between Thevenin and Norton resistance?
- What is the basis of differential and integral calculus?

NETWORK THEOREMS AND LAWS

The concepts of network theorems and laws are usually associated with a highly mathematical approach. Engineers use them to calculate values of circuit parameters and to design systems. As a technician you are primarily interested in making the systems work. However, it does help to be able to understand how some of these concepts are used for analyzing circuits.

Maximum Power Transfer Theorem. Figure 9-1 shows a simple dc power source. This could be a battery, or a dc generator, or any other source of dc voltage. We will represent all dc voltage sources as batteries in this discussion.

The battery voltage and the internal resistance of the battery are inseparable. In other words, there will always be a voltage and associated internal resistance. The question is, what value of R_L should be connected across the terminal of the battery so that it will receive maximum power?

The maximum power possible will always be delivered when R_L is equal to R_I. This is the maximum power transfer theorem. However, circuits are not always designed for maxi-

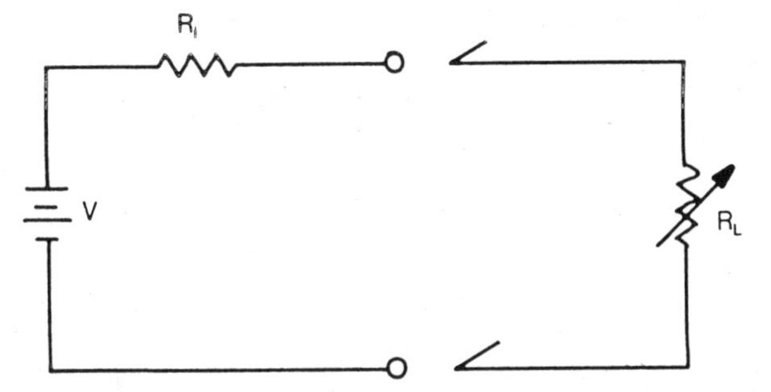

Fig. 9-1. This simple dc supply shows the voltage and internal resistance separately. Changes in R_L will change the terminal voltage because of the drop across R_l.

mum power transfer. The reason for this is that the efficiency of a circuit is also an important consideration in most applications. Efficiency can be calculated by the following equation:

$$\text{efficiency} = \text{power out/total power}$$

Total power is the internal and external power combined. Power out is the actual power delivered to the load resistance. If you multiply the right side of the equation by 100, you get percent efficiency. When maximum power is being transferred, the efficiency is only 50 percent.

If you add a switch to the circuit of Fig. 9-1, it could represent a simple flashlight. In that case, R_L would be the light bulb and V with R_l would be the battery. This is shown in Fig. 9-2. There are some problems, or tradeoffs, that you would face designing this very simple circuit. They were discussed in chapter 5 and are reviewed here.

- How long do you want the battery to last? If battery life is an important concern, you certainly do not want maximum power transfer, because it could give a very unsatisfactory battery life.

On the other hand, if brightness is your concern then you want to get as much power as possible to the light bulb, so your design would tend toward maximum power.

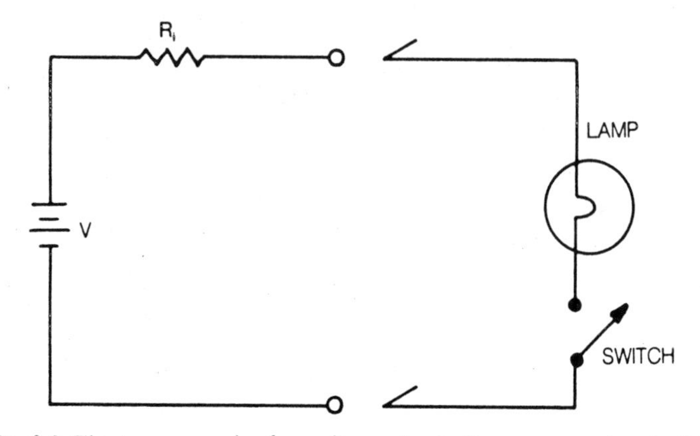

Fig. 9-2. This is an example of a nonlinear circuit. Current cannot be solved by Ohm's law.

- A second consideration is the effect of battery deterioration. If you desire the circuit in Fig. 9-2 to have maximum brightness when the battery is new, then certainly the brightness is going to decrease as the battery gets older. The reason is that there is a voltage drop across the internal resistance. This internal resistance increases as the battery gets older. So, there is a decrease in performance as the battery ages, which will affect the brightness. If brightness is the criterion for your design, you have to make a compromise between output power and battery life.
- A third problem is related to the light bulb. When the bulb is cold, it has one resistance and when it is warm it has another resistance value. Do you design this circuit for its hot resistance? What is the actual value of resistance for the amount of brightness that you are designing for?

You cannot solve for the value of current in the circuit of Fig. 9-2 by using Ohm's law. The light bulb resistance is nonlinear and Ohm's law will only work for linear devices. In other words, you do not know the value of the bulb resistance.

You can solve for the value of current when the bulb is hot and its hot resistance is known. However, the value of current you get will not be correct when the bulb is cold, at the start of operation. Also, it will not be correct when the flashlight is switched on and off at irregular intervals.

As the battery ages, the bulb temperature changes due to the lower current. That, in turn, changes the resistance of the bulb.

A relatively simple design problem was chosen for this book. However, the same tradeoffs are encountered in many power devices, and with different types of supplies.

Maximum AC Power. Maximum power transfer for ac circuits is somewhat more complicated than for dc. In order to get the maximum output power, the load impedance must be the conjugate of the internal impedance of the generator. This combination is shown in Fig. 9-3. When the internal impedance can be represented with a resistor and capacitor in series, then that internal impedance is normally written as $R\text{-}jX_C$.

In order to get maximum power transfer, you need an external inductor to resonate with the internal capacitance. The external impedance in this illustration is represented by $R+jX_L$. (See Fig. 9-3) The conjugate of $R-jX_C$ is the value $R+jX_L$.

In the same illustration, another internal impedance is represented by $R+jX_L$. The conjugate is $R-jX_C$. Putting it another way, you need a capacitor in the external circuit to resonate with the inductance.

At resonance, the phase angle between the voltage and current is zero degrees, and maximum power will then be obtained when the internal and external resistance values are equal.

Power factor is a measure of how nearly the voltage and current are in phase in an ac circuit. By definition:

$$\text{Power Factor} = \text{Cos}\ \phi$$

where ϕ is the phase angle between the voltage and current.

Phase angle is calculated by the following equation. Power factor has a value of 1.0 (100 percent) when voltage and current are in phase:

$$\phi = \tan^{-1} X_L/R$$

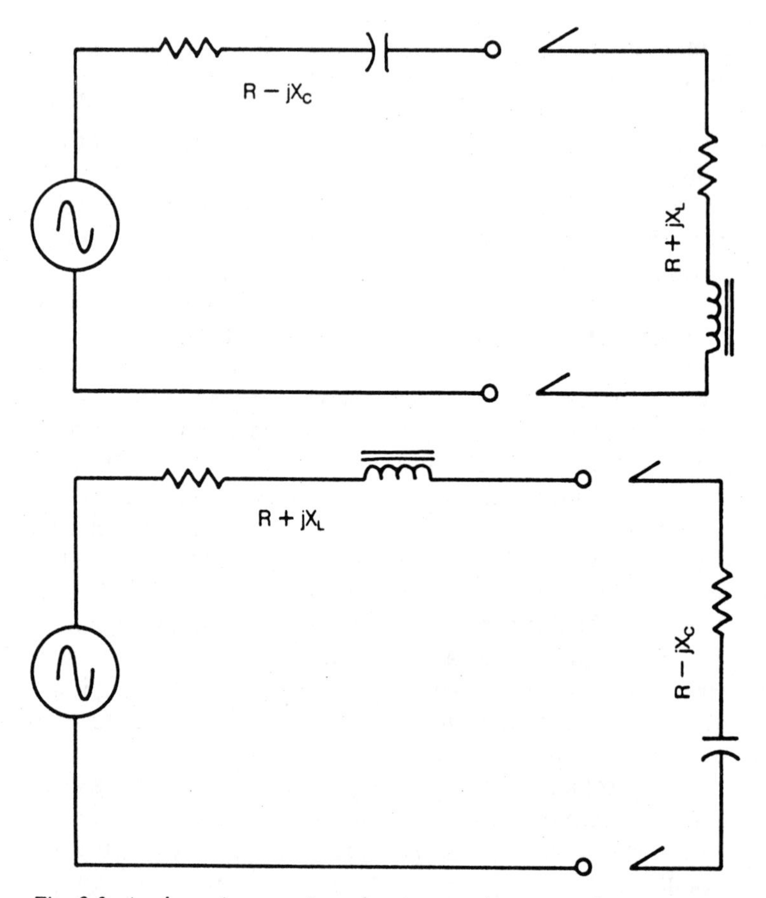

Fig. 9-3. As shown in a previous chapter, a conjugate match is needed for maximum power transfer.

or:

$$\phi = \tan^{-1} X_C/R$$

where ϕ is the phase angle, X_C or X_L is the equivalent inductive or capacitive reactance for the circuit, and R is the circuit resistance. Conjugate matches are very important in some applications of transmission lines and four-terminal networks.

Methods of Matching Impedances. Having briefly reviewed the theory of impedance matching and maximum power transfer, it is now possible to look at some practical circuits that accomplish this. Many impedance matching circuits are based on the principles of four-terminal networks.

Assume for a moment that you have a four-terminal network—that is, a two-terminal pair. As shown in Fig. 9-4, it consists of two input terminals and two output terminals. There is an internal combination of series and parallel resistance circuitry between the two terminals. It really does not matter how the resistors are connected. They can always be represented by the black box shown in Fig. 9-4. Assuming that this is a passive circuit, that is it has no internal sources of voltage, then the black box can be completely simulated by either of the two resistor circuits shown in Fig. 9-4.

Four parameters describe the black box:

- The input resistance with the output open: R_{AB}

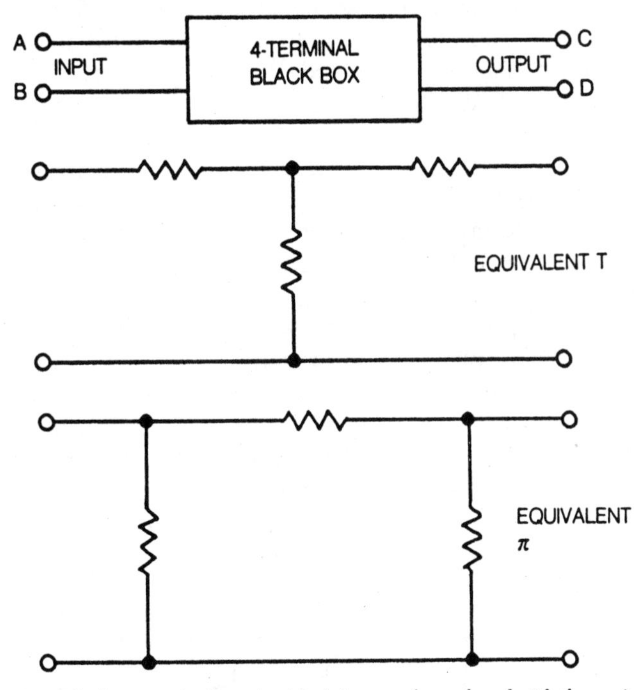

Fig. 9-4. The four-terminal passive black box can be replaced with the equivalent T or π. Measurements can only be made across the input and output pairs. For example, you cannot measure the resistance between A and C.

- The input resistance with the output shorted: R'_{AB}
- The output resistance with the input open: R_{CD}
- The output resistance with the input shorted: R'_{CD}

There are, then, four possible measurements to describe the black box. Any three of these are sufficient to find the equivalent circuit. The equivalent T circuit can completely replace the black box as far as the input and output parameters are concerned. So, if you make the same measurements on the equivalent T circuit as described for the black box, you will get exactly the same resistance values. Likewise, if you make the same measurements on the equivalent Π you will still get the same values. These measurements must only be made across input and output terminal pairs.

It should be apparent then that the black box can be used to match impedances even though the input and output resistances are not equal. If they are equal it is a symmetrical problem and very easy to solve. If they are not equal then the computation is more complicated. What is important to you as a technician is that these circuits completely describe the black box.

As an example of its operation, suppose you wished to match the high impedance circuit of a VFET to the low impedance of a two-terminal device such as a speaker. Remember that a VFET is a field effect transistor that can be used as a power amplifier.

This setup is illustrated in Fig.9-5. We will assume that the impedances of the amplifier and two-terminal device are different, but there is no phase angle between the voltage and current. Then, either the T or Π network can be connected between the two devices and the impedances can be matched. That means that the VFET amplifier will "believe" that it is looking into its own internal impedance, and the two-terminal device believes that it is looking into a generator that has an impedance equal to its impedance.

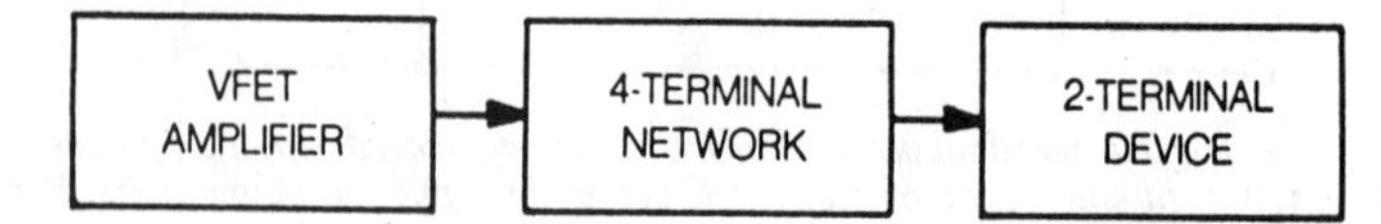

Fig. 9-5. This is an example of four-terminal network applications.

It is possible, through mathematical manipulation, to design the equivalent T or Π circuit so that it presents the minimum amount of loss between the two devices. In that case it is called a minimum insertion loss pad.

The four-terminal network can be designed to have a varying amount of loss so that the amount of power delivered to the two-terminal device can be adjusted without upsetting the impedance match. That kind of device is called an attenuator. You will see attenuators as the input circuits of oscilloscopes.

Even though a minimum insertion loss pad is used in the circuit for matching impedances there will still be some power dissipated in the four-terminal network. This is obvious because it is made with resistors, and current will flow through it when the output power is being received.

A transformer is another device that can be used for impedance matching. It has the advantage of not having resistance. More accurately, the resistance is minimal compared to the resistance of the pad or attenuator.

Figure 9-6 shows how the impedance ratio of a transformer is related to the turns ratio. An example of this application is matching an audio amplifier to a speaker. In that case it is called an "output transformer."

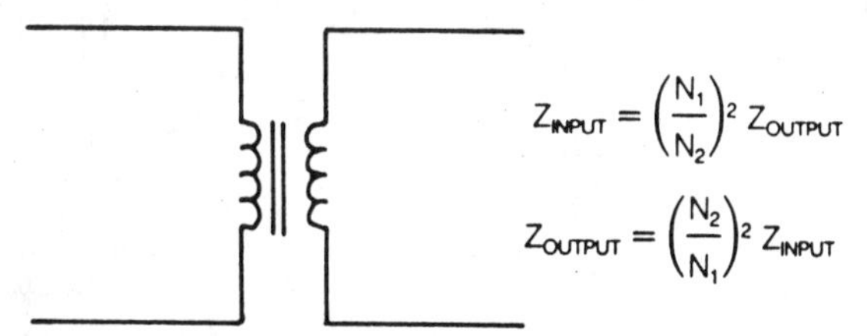

Fig. 9-6. By way of review, impedance matching by transformer is illustrated here.

The four-terminal networks described so far have pure resistance and it is assumed there is no phase angle between the voltage and the current. Another way of saying this is that there are no reactive components to be considered.

Another kind of passive four-terminal network is specifically designed for ac systems. It is called a filter. Filter circuits can be designed entirely on the basis of input and output impedances. Four kinds are illustrated in Fig. 9-7.

You should be able to recognize each of the filter types from the configuration of reactive components. Keep in mind that capacitors pass high frequencies and reject low frequencies and inductors pass low frequencies and reject high frequencies.

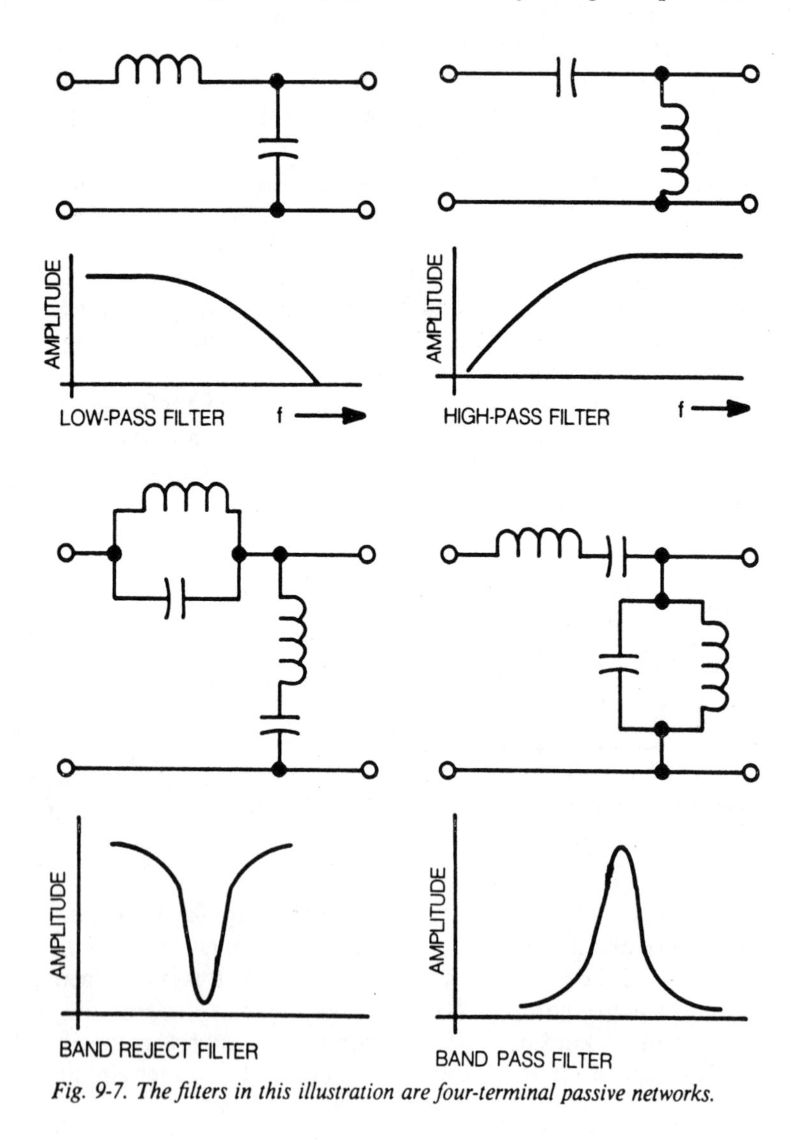

Fig. 9-7. The filters in this illustration are four-terminal passive networks.

From that, and the fact that parallel tuned circuits reject a given frequency, whereas series tuned circuits pass only the resonant frequency, you can usually determine the type of filter from a study of the configuration.

Sometimes when you encounter filters in a circuit, they will be given unusual names like Chebyshev, Butterworth, Constant-k, M-derived, etc. The names have nothing to do with the input and output impedance or frequency response. These are names of the designers of the original circuit, or as in the case of constant k and M derived, they are simply an indication of a procedure for designing filters.

The most commonly encountered filter is the low-pass type found at the output of a rectifier. Low-pass filters are also encountered in phase-locked loops.

Active Four-Terminal Networks. Active four-terminal networks have one or more sources of voltage inside the black box in addition to passive components. This completely changes the characteristics of four-terminal networks. Now, it is necessary to measure input voltages and corresponding currents. Four useful parameters are:

- The output voltage with the output terminals open,
- The output current with the output terminals short circuited,
- The input voltage with the input terminals open,
- The input current with the input terminals short circuited.

Using combinations of these parameters, it is possible to completely describe the black box with simple T and Π networks, provided you include sources of voltage. There are different methods of describing active four-terminal networks on the basis of types of parameters chosen. For example, you could completely describe it with voltage parameters or current parameters, or with Z (impedance) parameters. Engineering books give long but precise equations for identifying circuits on the basis of those choices.

An important choice is a combination of voltage and current parameters. These are called hybrid or *h* parameters. You will see these parameters used to define bipolar transistors. For example H_{FE} and H_{FB} are forward transfer parameters with the emitter or base connected as a common between the input and output terminals. Using the h parameters, which the manufacturers of transistors provide, you can completely design a transistor amplifier without having ever taken into consideration the transistor itself. It is simply represented by an equivalent four-terminal circuit.

Today, computer programs permit you to feed in the parameters that have been supplied by manufacturers. Using those parameters, the computer generates a very useful circuit design. As an example, the design of filter circuits can now be done with a software package.

Thevenin's and Norton's Theorems. Unlike other network theorems and laws, Thevenin's and Norton's theorems are based upon imaginary generators. Although they do not exist in the real world, they are very useful for analyzing circuits.

The first imaginary concept to consider is the constant-voltage generator. A constant-voltage generator delivers the same amount of voltage regardless of how much current is flowing through it.

The only way you can have a constant-voltage generator is to have no internal resistance, which, of course, is impossible. All voltage generators have internal resistance.

The second imaginary concept is the constant-current generator. It delivers the same amount of current regardless of the amount of load resistance. In fact, it will continue to deliver that current when the output terminals are short-circuited.

Figure 9-8 shows the Thevenin (constant-voltage) and Norton (constant-current) generators connected to an imaginary internal resistance, R_{TH} and R_N. This is the way they are most often represented.

Consider, now, a two-terminal active network in a black box as shown in Fig. 9-9. It is necessary to assume that there is a combination of two-terminal, linear, bilateral circuit elements connected inside. A linear component is one that complies with Ohm's law. If you double the voltage across it, the current through it will double. A bilateral component is one that conducts current equally in either direction. Since the black box in Fig. 9-9 is active, it follows that there is at least one voltage source inside.

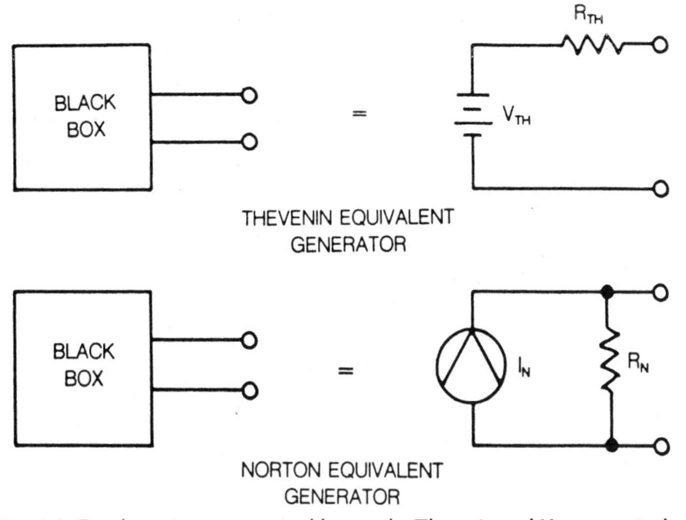

Fig. 9-8. For the active two-terminal boxes, the Thevenin and Norton equivalent generators are shown.

If we consider the circuit inside the black box to be dc only, then it follows that the two-terminal components are linear resistors and the dc voltage sources can be represented by batteries.

The overall result is that the black box can be replaced by either of the equivalent circuits shown in Fig. 9-9. If you look into the terminals of the Thevenin generator you will see a voltage source and a resistance of R_{th}. Since the Thevenin and Norton generators are both equivalent to the two-terminal active circuit, it follows that they are also equivalent to each other. So, when you look into the two terminals of the Norton generator, you will also see R_{th}. In other words, $R_{th} = R_n$.

The only way you could see R_{th} (or, R_n) looking into the two terminals of the Norton generator is if the internal resistance of the Norton generator were infinitely large.

To summarize, the internal resistance of V_{th} must be zero ohms, and the internal resistance of I_n must be infinitely high. No matter how many components and voltage sources are inside the black box, it can *always* be represented by either of the generators shown in Fig. 9-9.

This is a very important network concept. It permits cur-

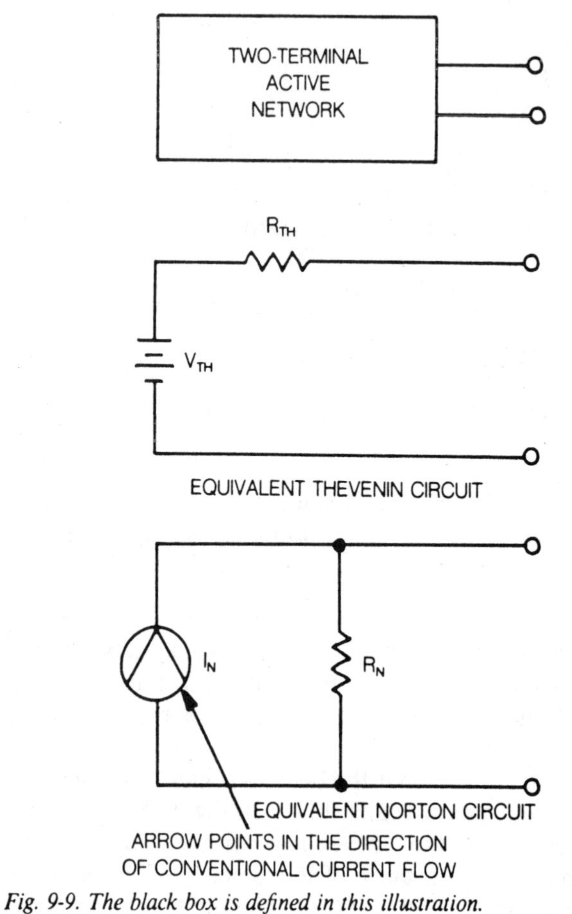

Fig. 9-9. The black box is defined in this illustration.

rents of nonlinear components to be determined in circuits where Ohm's law cannot be directly applied. The best way to show how these circuit ideas can be used is to use them in a simple solution.

PROBLEM: How much current will flow through the diode in the circuit of Fig. 9-10?

SOLUTION: The first step is to remove the diode. This results in a two-terminal active network as shown in Fig. 9-11. As with all active two-terminal networks containing linear circuit elements and one or more sources of voltage, the network

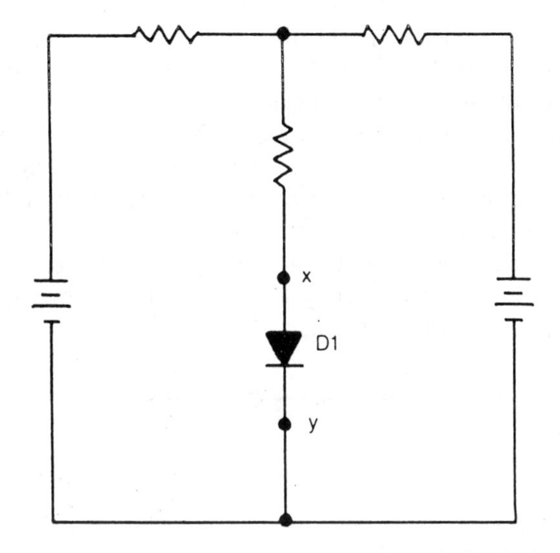

Fig. 9-10. This is a nonlinear circuit. Current through the diode cannot be determined by Ohm's law.

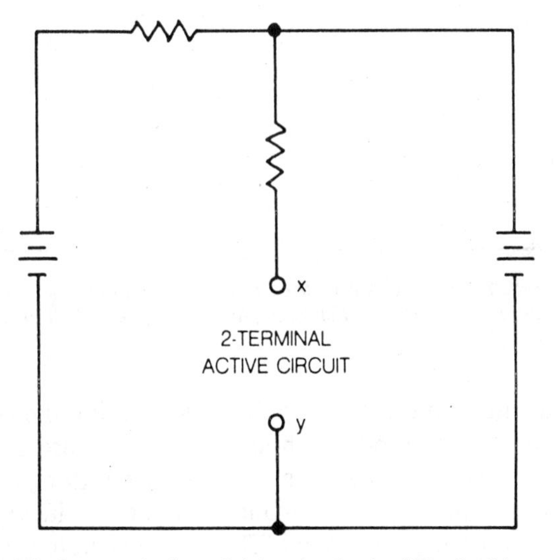

Fig. 9-11. The first step in theveninizing the circuit of Fig. 9-10 is to remove the nonlinear component.

in Fig. 9-10 can be replaced with a Thevenin (or Norton) generator as shown in Fig. 9-12. The values of V_{th} and R_{th} must be calculated in order to make the original circuit equivalent to the Thevenin circuit. The required calculations will not be done here because we are interested only in the meaning of the theorems at this time. However, a method of measuring V_{th} and R_{th} will be discussed in the next section.

If the diode is connected across the Thevenin generator of Fig. 9-12 it will have exactly the same current through it as when connected into the original circuit.

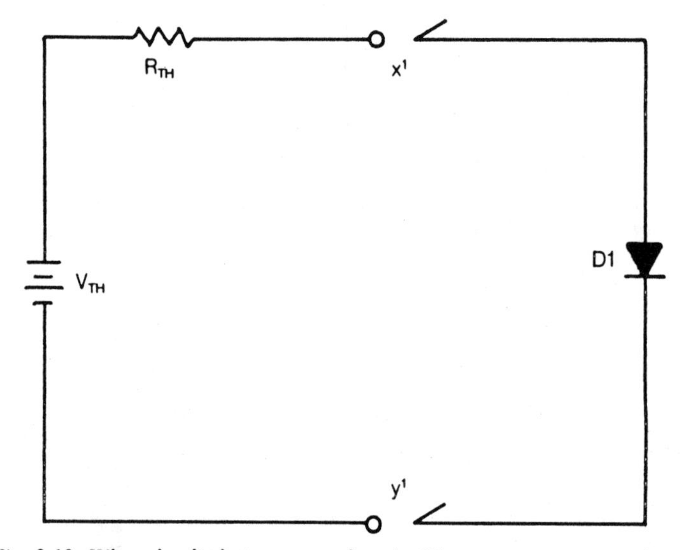

Fig. 9-12. When the diode is connected to the Thevenin circuit, a number of different solutions are possible. For example, the diode equation or a load line solution can be made.

You can calculate the current flow by using the equation for semiconductor diodes. Another way is to use a graphic solution by drawing a load line. The important thing to understand is that a solution can be made when the nonlinear diode is connected into the equivalent Thevenin generator. In other cases, the Norton generator will permit the solution to a nonlinear problem.

If you have ac generators instead of batteries, and impedances instead of resistances, you have the ac version of Thevenin and Norton generators.

Practical Thevenin and Norton Generators. The idea behind Thevenin's and Norton's theorems can be put to use without the need for mathematical calculations. As an example, consider the circuit in Fig. 9-13. Assume that you want to know what value R_x must have to receive maximum power from the circuit, and, you want to know the value of maximum power.

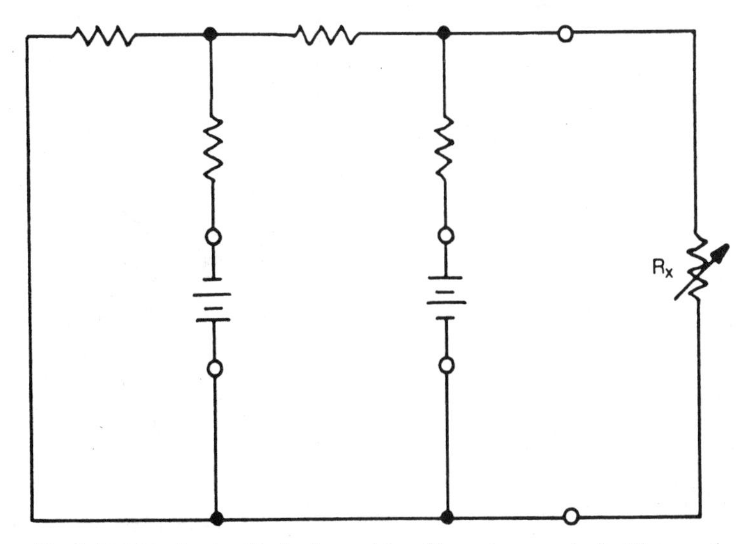

Fig. 9-13. This circuit will also be used in a Thevenin example. In this case, the Thevenin generator will be obtained by measurement.

Figure 9-14 shows the procedure for measuring V_{th}. You will need a high-impedance voltmeter like the popular digital voltmeters used today, or you can use an oscilloscope to make the voltage measurement.

As shown in Fig. 9-14, remove R_x from the circuit and measure V_{th}. Then, replace the voltage sources and measure R_{th}, also shown in Fig. 9-14. Once you have drawn the equivalent Thevenin generator with the measured values, set R_x equal to R_{th} to obtain maximum power. The value of that power is easily obtained by calculating the total power—that is, the power dissipated by both R_{th} and R_x—and dividing that value by 2, because only half of the total power is dissipated by R_x.

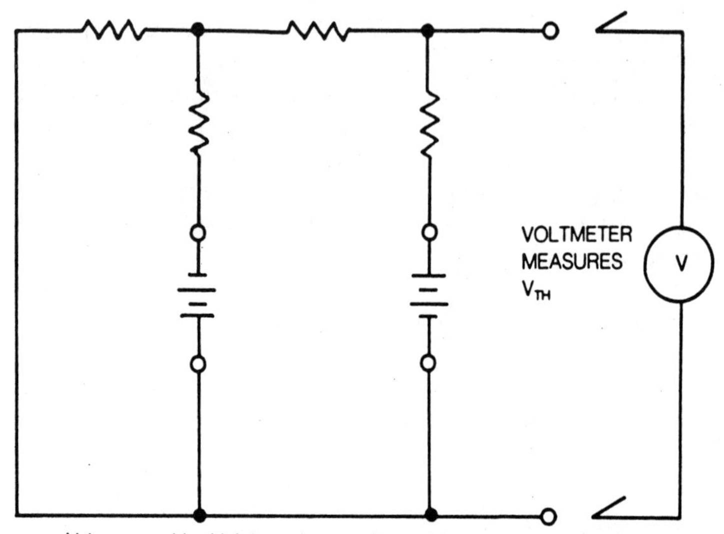

V (measured by high-impedance voltmeter) is approximate value of R_{TH}.

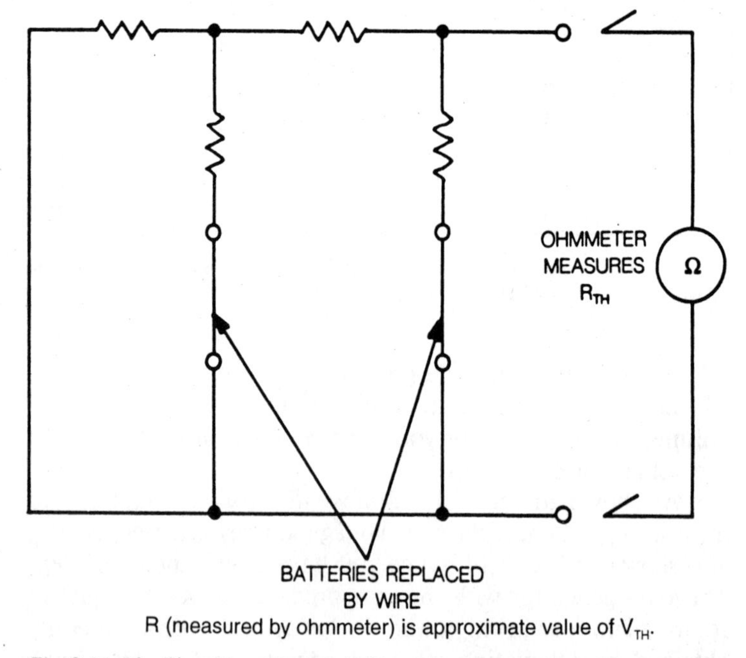

R (measured by ohmmeter) is approximate value of V_{TH}.

Fig. 9-14. The Thevenin voltage and Thevenin resistance are obtained as shown here.

REVIEW AND EXTENSION OF SOME BASIC CIRCUIT IDEAS

One subject normally covered in a basic electronics course is time constants. They are related to the basic RL and RC circuits shown in Figs. 9-15 and 9-16.

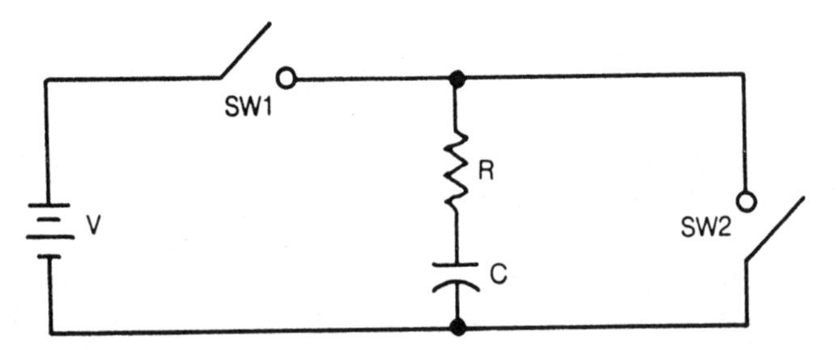

Fig. 9-15. Transient voltages and currents can be obtained by operating SW1 and Sw2.

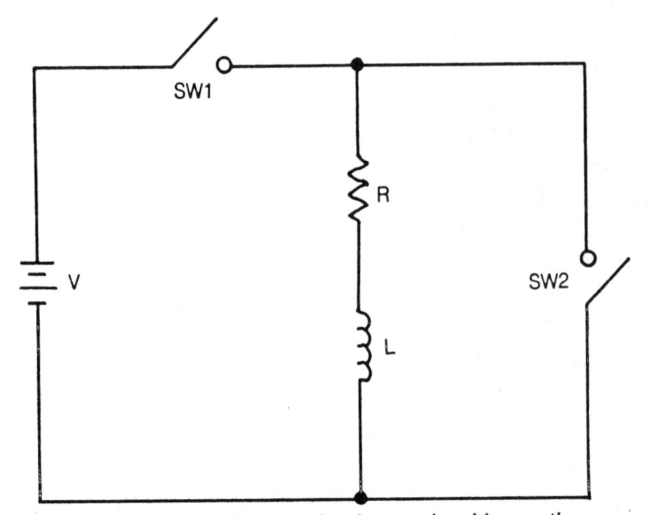

Fig. 9-16. The capacitor in Fig. 9-15 has been replaced by a coil.

Consider the RC time constant circuit. With SW2 closed, the capacitor will eventually become totally discharged. This is the starting point of the time constant story.

When SW2 is open and SW1 is closed, the capacitor will charge through the resistor. The time it takes for it to charge to 63 percent of the applied voltage is called one time constant (T).

By definition, five time constants are required to charge the capacitor to the applied voltage.

The equation for the time constant is:

$$T = RC$$

$$\text{and, Full Charge} = 5RC$$

where T is in seconds,

R is in Ohms, and
C is in farads.

At the end of five time constants, the capacitor is considered to be fully charged or discharged. Although this is not precisely true if you go by the mathematics of time constants, it is close enough for practical work. Five time constants is also considered to be the time required for an RL circuit to reach maximum current or zero current.

Assume that the capacitor has become fully charged and switch SW1 is opened. Then, closing SW2 will cause the capacitor to discharge through the resistor. The time that it takes for the voltage across the capacitor to drop to 37 percent of the original applied voltage is also called one time constant (T).

The case of the RL time constant circuit is slightly more complicated. In the first place, the countervoltage produced by the inductor when there is a change in current through it can be sufficiently high to produce a voltage arc at the switch. When this happens, time constant equations are no longer applicable.

Here, we will assume that the countervoltage is not sufficient to produce the arcing. When SW2 is open and SW1 is closed, current will begin to flow through the inductor. You will remember that an inductor opposes any change in current through it. Remember also that a capacitor opposes any change in voltage across it. Therefore, the current in the circuit cannot reach its maximum value instantly. It must overcome the countervoltage produced by the coil.

The current builds up in much the same way as the voltage across the capacitor builds up and reaches 63 percent of its maximum value at the end of one time constant. If SW1 is open and SW2 is closed, the energy stored in the coil will be returned to the coil and a discharge current will flow. It will take one time constant for the current to drop to 37 percent of its maximum value. The equation for the LR time constant is:

$$T = L/R$$

For full current or zero current: $T = 5L/R$

where T is the time constant in seconds,
L is in henries, and
R is in ohms.

A universal time constant curve can be used to determine the actual value of charging or discharging voltage or current. Figure 9-17 shows the curves. In order to use the curves, it is necessary to divide the time constant axis into seconds, microseconds, picoseconds, or whatever is applicable.

PROBLEM: In the circuit of Fig. 9-18 the resistance value is 10K and the capacitance value is one microfarad. What is the value of one time constant, and how long will it take the capacitor to charge to 90 percent of the applied voltage?

SOLUTION: The time constant is obtained as

$$T = RC$$

$$= (10 \times 10^3) \text{ ohms X } (1 \times 10^{-6}) \text{ microfarads}$$

$$= 0.01 \text{ second}$$

The time constant curve of Fig. 9-19 has been marked to represent this value. Note that each of the time constant segments is equal to the value obtained in the calculation. In each case it is added to the previous value.

Having marked the time constant axis, you can determine how long it takes to get to 90 percent of the value by entering the curve at the 90 percent mark as shown by the heavy line. Move to the charging curve from the 90 percent point and then down to the time constant axis. If you want good accuracy, you will have to divide the time constant segments into smaller units, but you can estimate the value for most applications.

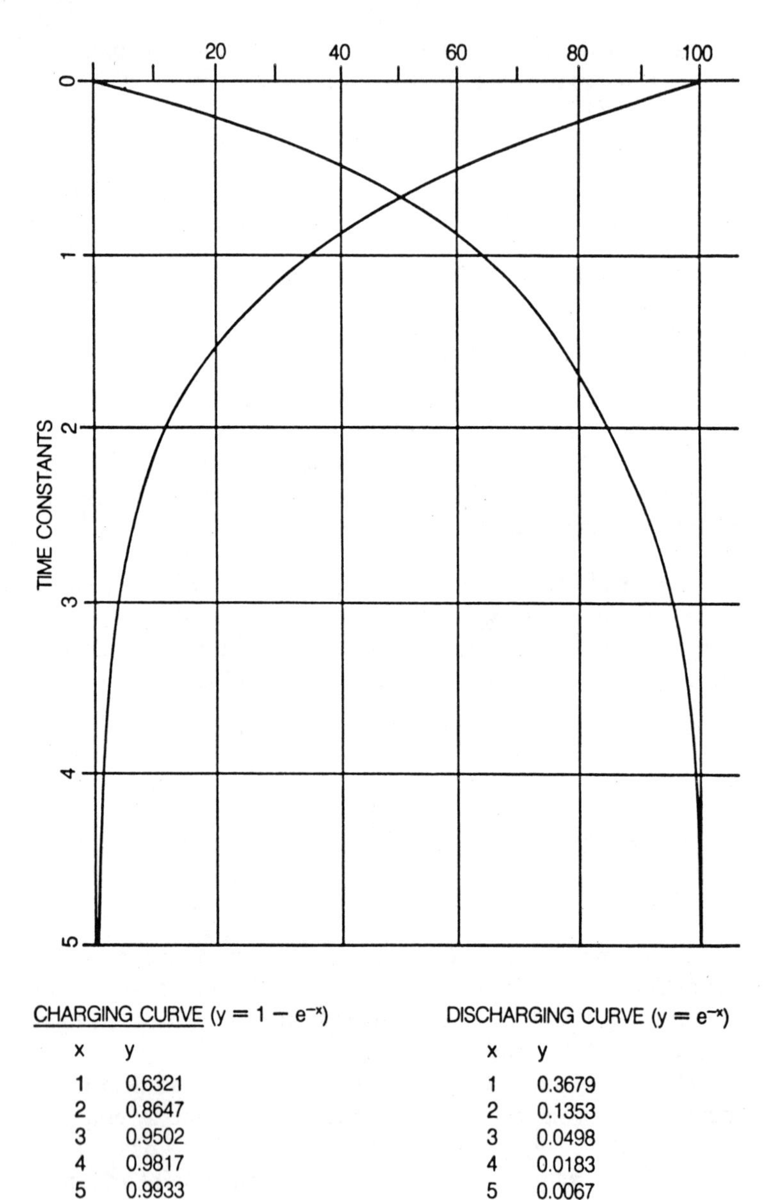

CHARGING CURVE ($y = 1 - e^{-x}$)

x	y
1	0.6321
2	0.8647
3	0.9502
4	0.9817
5	0.9933

DISCHARGING CURVE ($y = e^{-x}$)

x	y
1	0.3679
2	0.1353
3	0.0498
4	0.0183
5	0.0067

Fig. 9-17. The universal time constant curves are illustrated in this graph.

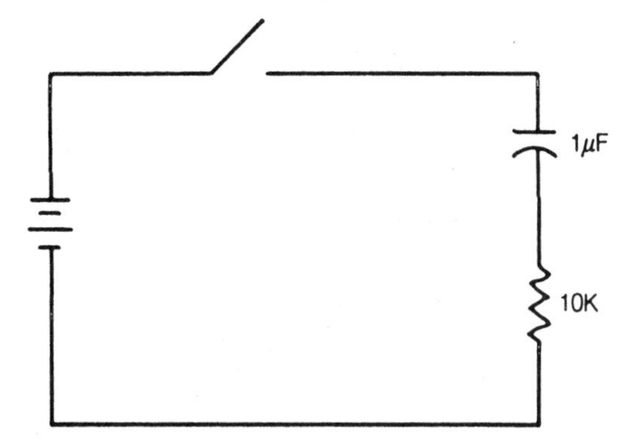

Fig. 9-18. The values used in this circuit are for a practice problem.

The value of t is shown graphically to be about one-third of the way from 0.02 to 0.03 seconds. Call it 0.03. So, the time required is 0.023 seconds. Modern calculators make it easy to determine the times more accurately by using the exponential equations. They are listed here:

Voltage across a charging capacitor:

$$V_C = V - V/E^{t/RC}$$

* Current growth in an inductor:

$$i_L = I - I/E^{tR/L}$$

Voltage across a discharging capacitor:

$$V_C = V/E^{t/RC}$$

* Current decrease in an inductor:

$$i_L = I/E^{tR/L}$$

* Due to countervoltage generated by switching RL circuits, these equations may not work.

where E is a constant equal to 2.71828 (e on calculators).

The time for charging to 90 percent of the applied voltage in the previous problem can now be calculated more precisely. It is necessary to solve the time constant equation for charging

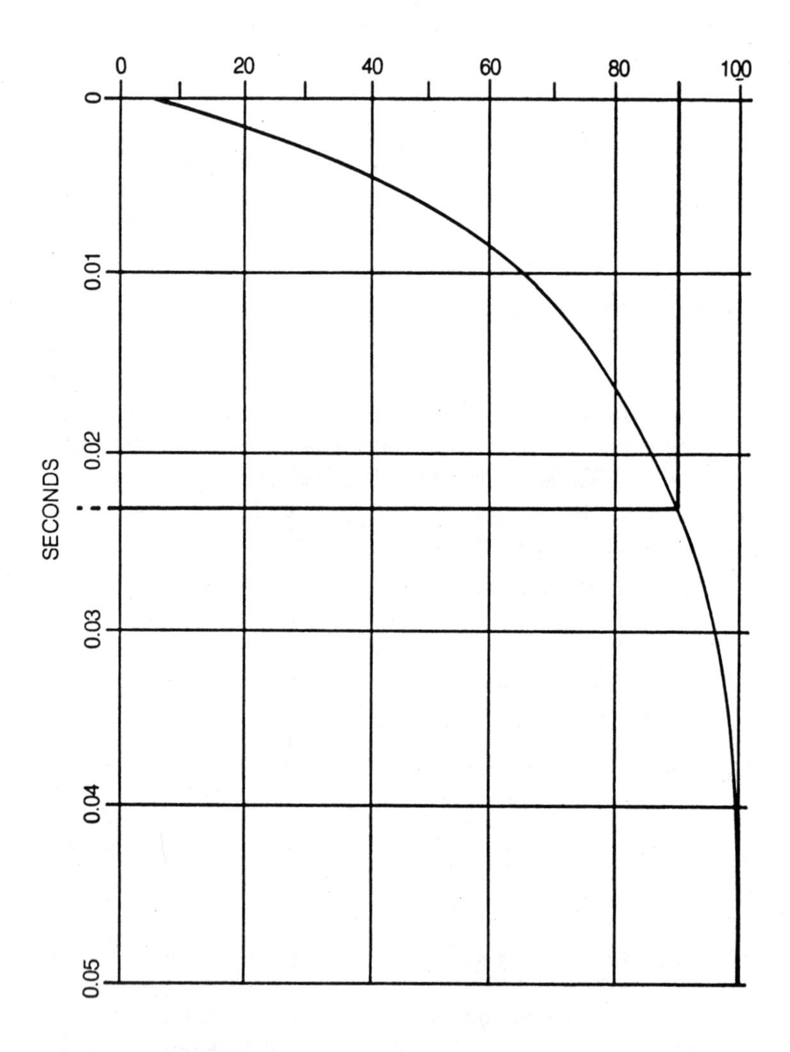

Fig. 9-19. The graphical solution is shown here.

to get the value of T. The procedure is shown in Fig. 9-20 for technicians who like their theory laced with a bit of mathematics.

$$V_C = V\left(1 - \frac{1}{e^{t/RC}}\right)$$

$$-V_C = -V + \frac{V}{e^{t/RC}}$$

$$\frac{V - V_C}{V} = \frac{1}{e^{t/RC}}$$

$$e^{t/RC} = \frac{V}{V - V_C}$$

$$e^{t/RC} = 10$$

$$\frac{t}{RC} \ln e = \ln 10$$

$$\frac{t}{RC} = 2.30258 \text{ (note: RC = 0.01)}$$

$$t = \underline{0.023}$$

Fig. 9-20. The mathematical solution is given in detail in this illustration.

The basic time constant equations (T = *RC* and T = *L/R*) tell some important things about RC and RL circuits. For example, it takes longer to charge a capacitor through a resistor if you increase the resistance value, or increase the capacitance value, or both. In an *RL* circuit it takes longer for the current to reach its maximum value if you increase *L* or decrease *R*.

The charge and discharge of current and voltage follows a logarithmic curve in both RL and RC circuits. Circuits that employ time constants for producing sawtooth voltages will not produce a linear sawtooth unless some special provision is made.

The UJT Time Constant Circuit. One of the simplest nonsinusoidal oscillators, usually called relaxation oscillators, is shown in Fig. 9-21. It is a simple UJT oscillator. Its frequency is directly related to the RC time constant. A unijunction transistor (UJT) will not conduct until its emitter voltage reaches a certain percentage or decimal value of the applied voltage. The parameter that determines what percent value the emitter voltage has to reach before UJT conduction takes place is called the intrinsic standoff ratio.

Different UJTs have different values of intrinsic standoff ratio. The intrinsic standoff ratio is set by the manufacturer. One very logical value is 0.63. That means the UJT will conduct when the voltage across the capacitor is 63 percent of the applied voltage. At that point the emitter voltage will be at the right value to start the UJT into electron conduction from base 2 to emitter. In other words, the transistor will conduct at the end of one RC time constant.

The solid arrows in Fig. 9-21 show the charge path for the capacitor through the resistor to the power supply voltage. When the UJT conducts, it discharges the capacitor in a path shown by the broken arrows. The waveform at the emitter is a sawtooth, as shown in Fig. 9-21. Also, a positive pulse output can be obtained from base 2. This pulse occurs every time the discharge current flows through the resistor.

The curvature of the sawtooth in Fig. 9-21 can be undesirable in some applications. To get around that problem, a constant-current diode can be added to the circuit as shown in Fig. 9-22. The constant-current diode causes the capacitor to charge in a linear fashion instead of along the time constant curve as previously shown. This gives a linear sawtooth output that is more useful in many applications.

A constant-current diode can be made with a JFET as shown in Fig. 9-23. Constant current does not occur until the current through the JFET reaches a certain minimum value.

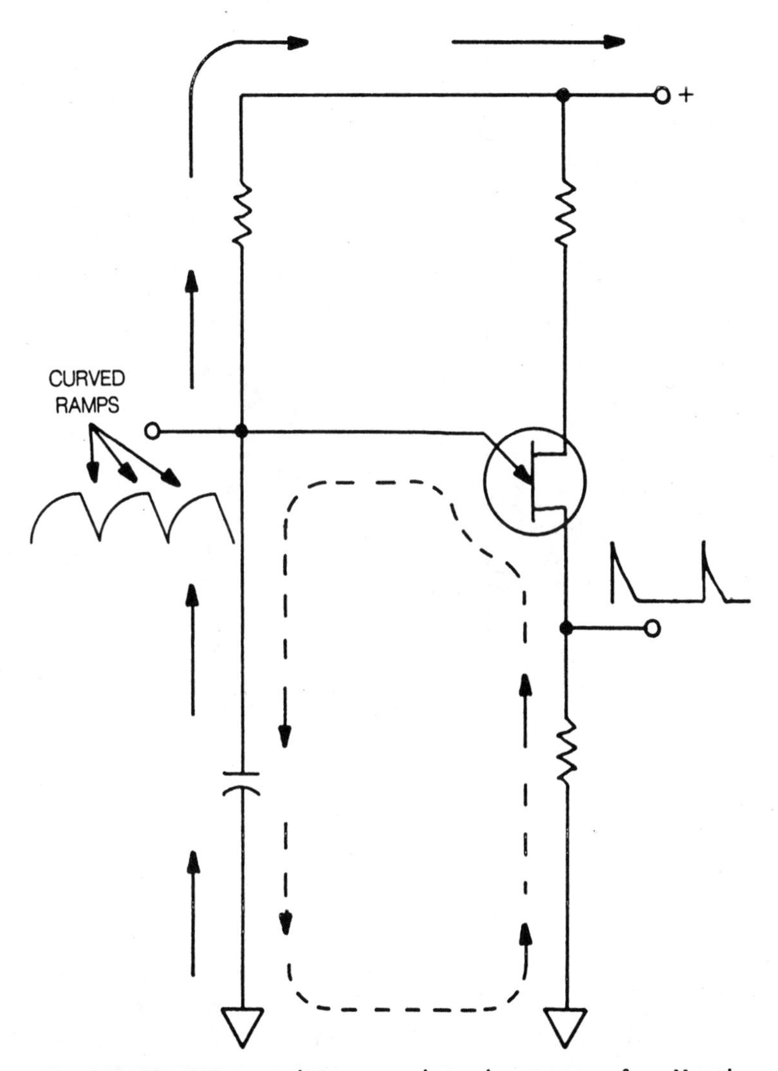

Fig. 9-21. The UJT can produce a sawtooth or pulse output waveform. Note the curvature in the sawtooth waveform.

However, as the beginning of the sawtooth along the time constant curve is relatively linear, this does not cause a serious problem.

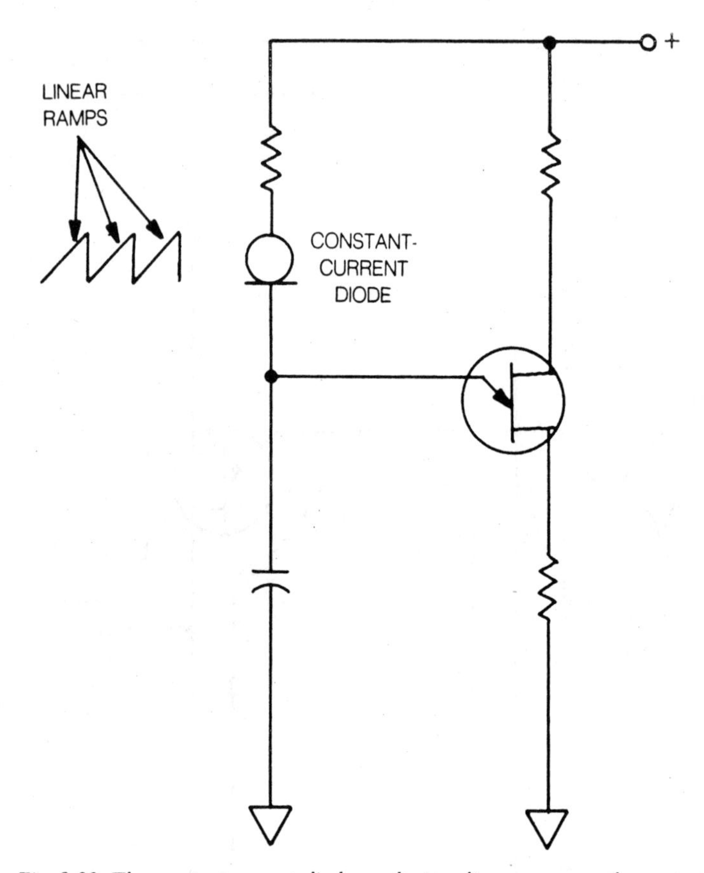

Fig. 9-22. The constant-current diode results in a linear ramp on the sawtooth waveform.

The Integrator and Differentiator. In chapter 2 the use of op amps for making an integrator and a differentiator was introduced. Figure 2-20 shows examples of the basic circuits. The concepts of integrators and differentiators will be extended next.

In mathematics, integration is the procedure for adding quantities. It is often represented by the operator, or symbol Σ. For example, the average value of any voltage waveform is equal to the area between the waveform and zero volts. You can see this represented in Fig. 9-24.

If the area is irregular, you can divide it into small rectan-

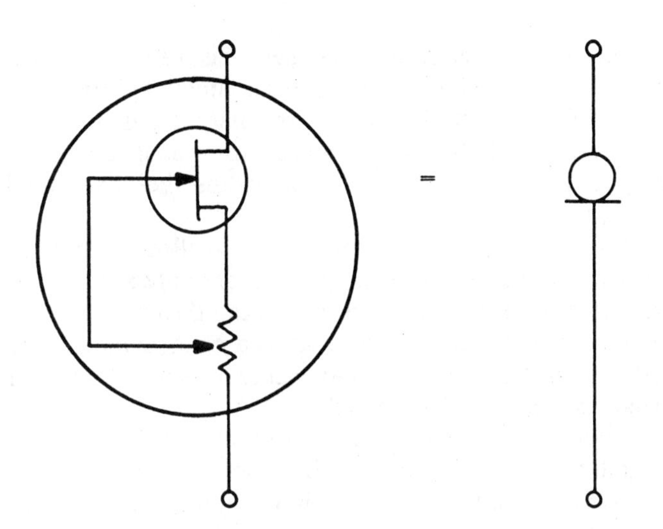

Fig. 9-23. Constant-current diodes are made this way. In a manufactured constant-current diode, the resistor is not variable.

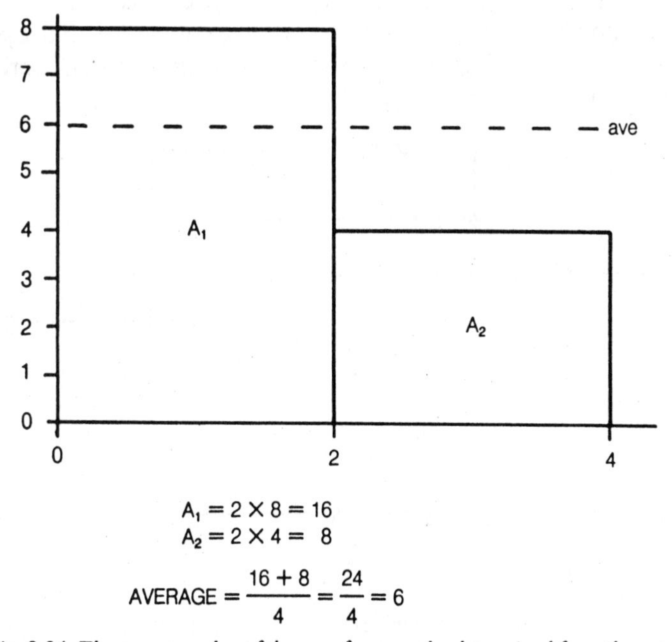

Fig. 9-24. The average value of the waveform can be determined from the areas under the curves.

gles and add the areas of the rectangles. This is also illustrated in Fig. 9-24. There will always be a certain amount of error when finding the area when the curve is irregular, because the rectangles do not exactly fit, as you can see from the illustration. By using rectangles with very narrow widths the error can be disregarded.

A special mathematical procedure—called integration—makes it possible to use rectangles that are only a single point wide. An infinite number of rectangles can then be used to find the exact area. Although integration is a mathematical trick, it produces exact area values. Exact values of average voltage and current can then be determined.

If you supply a large number of rectangular pulses to an integrating circuit, it will add the pulses. This is shown in Fig. 2-20. Actually, the circuit shown in the illustration does not produce perfect integration, but its output is very nearly equal to the sum of the rectangles.

In electronic circuits that use integrators, a specific number of pulses is added. When the output reaches a certain value, corresponding to a given number of pulses, it is used to trigger another circuit. Figure 9-25 shows an integrator that works better than the one in Fig. 2-20.

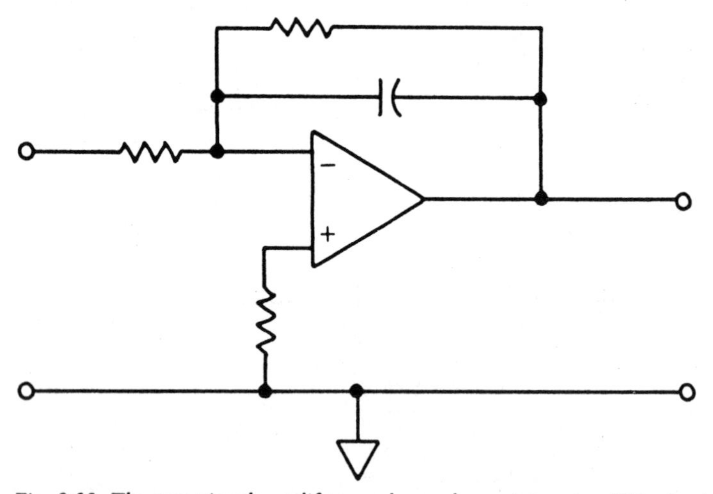

Fig. 9-25. The operational amplifier can be used as a integrator. This circuit works better than the one shown in a previous chapter. The capacitor serves to speed up the feedback voltage.

Differentiation in mathematics is the mathematical procedure used to find an instantaneous rate of change. It answers the question "how fast is one quantity changing with respect to another?"

Voltage induced in a coil at any instant can be determined by the equation:

$$v = -N\, d\phi/dt$$

In this equation, v is the instantaneous induced voltage and N is the number of turns of wire in the inductor. The negative sign signifies that it is a countervoltage. The expression $d\phi/dt$ means the rate of change of flux (ϕ). In other words, it represents how fast the flux is changing with respect to time.

You can determine the rate of change of any waveform with respect to time by drawing a tangent to the curve at the point of interest. Two examples are shown for the sine wave in Fig. 9-26.

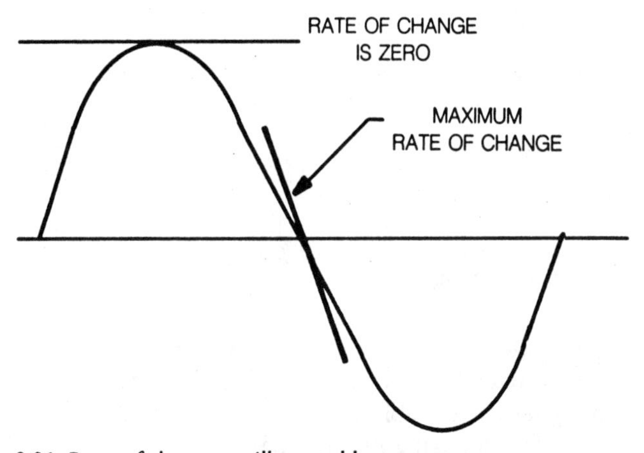

Fig. 9-26. Rates of change are illustrated by tangents to curves.

The output of a differentiating circuit is directly related to the rate of change of the input waveform. This is shown in Fig. 9-27. The circuit does not produce a 100 percent accurate differentiation, but the output is very nearly correct if the time constant of the RC combination is a very low value.

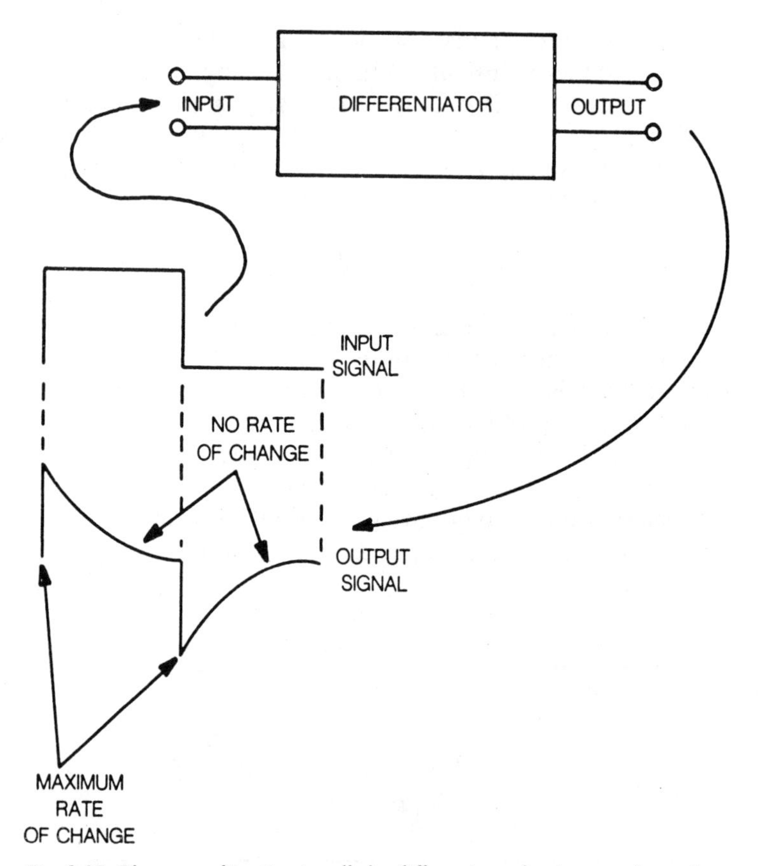

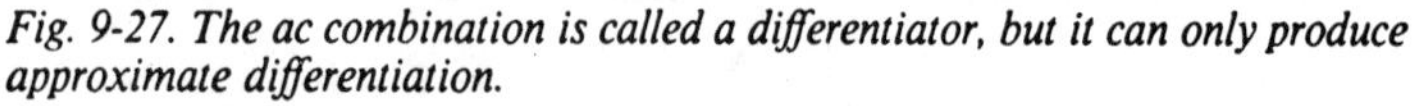

Fig. 9-27. The ac combination is called a differentiator, but it can only produce approximate differentiation.

Figure 9-28 shows the op amp version. This circuit will give better results than the one in Fig. 2-20. The added feedback capacitor makes it possible for the op amp to respond more rapidly to a fast-changing input signal.

Op Amp Buffers. If the output of an integrating or differentiating circuit is connected to a circuit that has capacitance or inductance, the shape of the waveform will be changed. This will defeat the purpose of the circuit in most cases.

Whenever it is necessary to isolate the output of one circuit from another circuit, a buffer can be used. Figure 9-29 shows an op amp buffer. It has a gain of 1.0 and a very broad bandwidth.

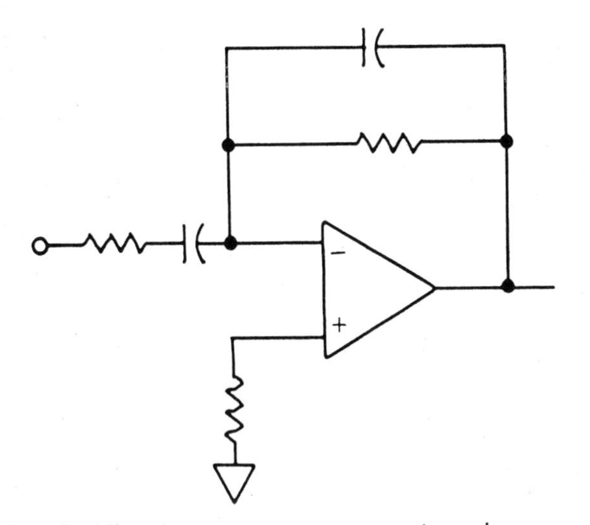

Fig. 9-28. The differentiator converts a square wave into pulses.

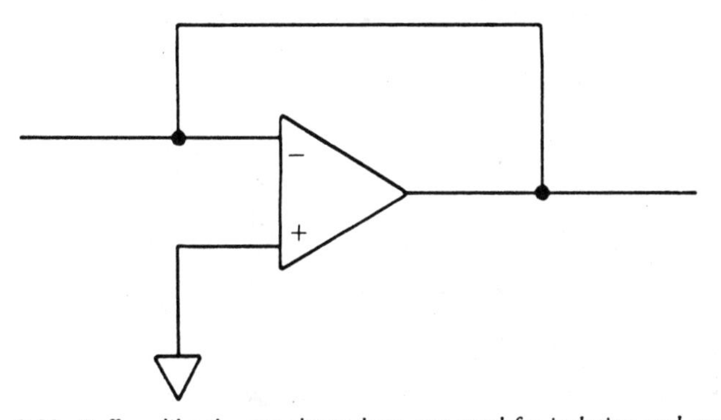

Fig. 9-29. Buffers, like the one shown here, are used for isolation and current gain.

Buffers are also used between oscillators and their output system to prevent them from being pulled off frequency by capacitance or inductance in that system.

The 555 Timer. You have no doubt worked with a 555 timer in your first-year electronics courses. It is a very versatile integrated circuit. If you are not familiar with this device you should review it periodically until you understand its operation. Other IC timers that perform similar operations are better understood by starting with the 555 timer.

Figure 9-30 shows two connections for the 555. In the astable multivibrator condition it produces a continuous output rectangular waveform. Unless special provisions are made, this waveform is not a square wave. The reason for this is the capacitor charge and discharge circuits are different.

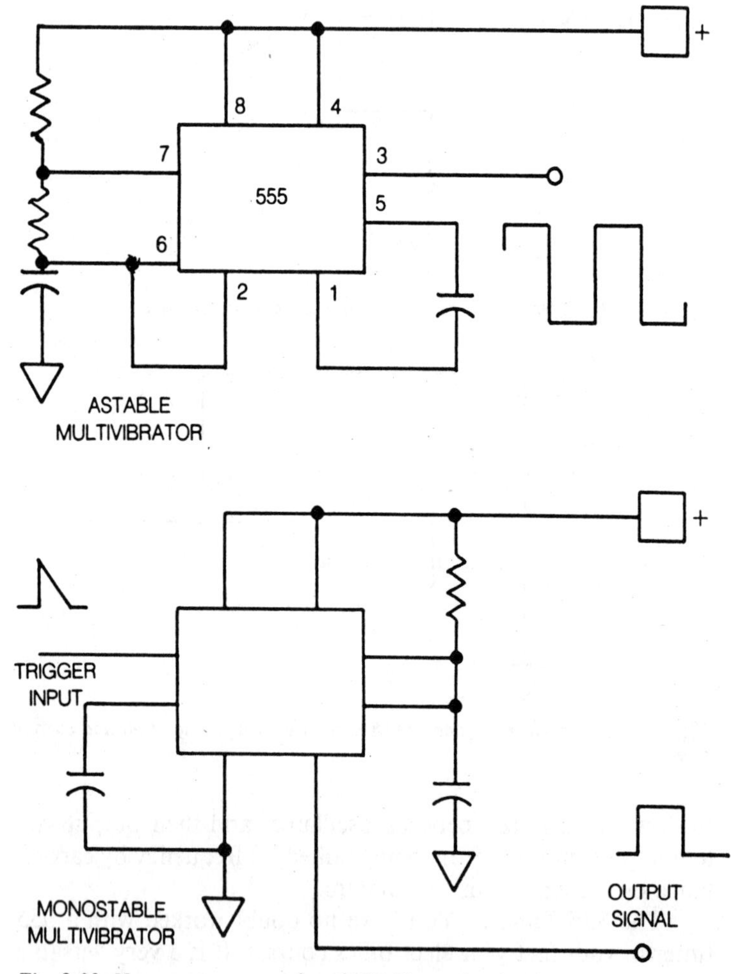

Fig. 9-30. Here are two examples of 555 Timer circuits.

A monostable multivibrator is sometimes called a pulse stretcher. It is normally in its stable condition until a short-duration trigger pulse arrives. That causes it to cycle through one flip flop operation and produce a rectangular output pulse. Because the output pulse is wider than the input trigger pulse, the device gets the name pulse stretcher.

This circuit is also used to shape waveforms. For example, the short pulse out of the UJT oscillator in Fig. 9-21 would not be satisfactory for some operations. It is not rectangular. By using that pulse to trigger a monostable multivibrator, a well shaped rectangular pulse can be obtained. (See Fig. 9-30.)

Monostable and astable multivibrators can be made with discrete components, but the integrated circuit versions are more convenient and occupy less "real estate"—that is, less space on the surface of a PC board.

Frequency Synthesizers. A frequency synthesizer produces a frequency that is dependent upon two different input frequencies. The phase-locked loop version is shown in Fig. 9-31. The blocks basic to all phase-locked loops are the phase comparator Φ, a low-pass filter, an amplifier (not always included), and a voltage-controlled oscillator VCO.

This basic phase-locked loop is modified to produce an output frequency (F_{out}) that is obtained by manipulating the divide-by-N1 and divide-by-N2 blocks.

Suppose for example that the divide-by-N1 and divide-by-N2 circuits are both set to divide by one. In other words, they produce no change in the crystal oscillator output frequency or in the frequency from the VCO. When those two frequencies are identical, there is a fixed dc output from the phase-locked amplifier circuit and the VCO is on center frequency. This frequency is equal to the crystal frequency.

Assume now that N2 is changed to divide by 2. In order to get a phase lock, the VCO must have a frequency that is twice the crystal frequency. So, after it is divided by 2, it is equal to the crystal frequency. By setting N2 to different values, you can obtain a wide range of frequencies that are crystal controlled.

Now, assume that N1 is set to divide-by-2 and N2 is set to divide-by-3. With N2 set to divide-by-3, VCO will have a frequency three times the input at the phase comparator circuit in

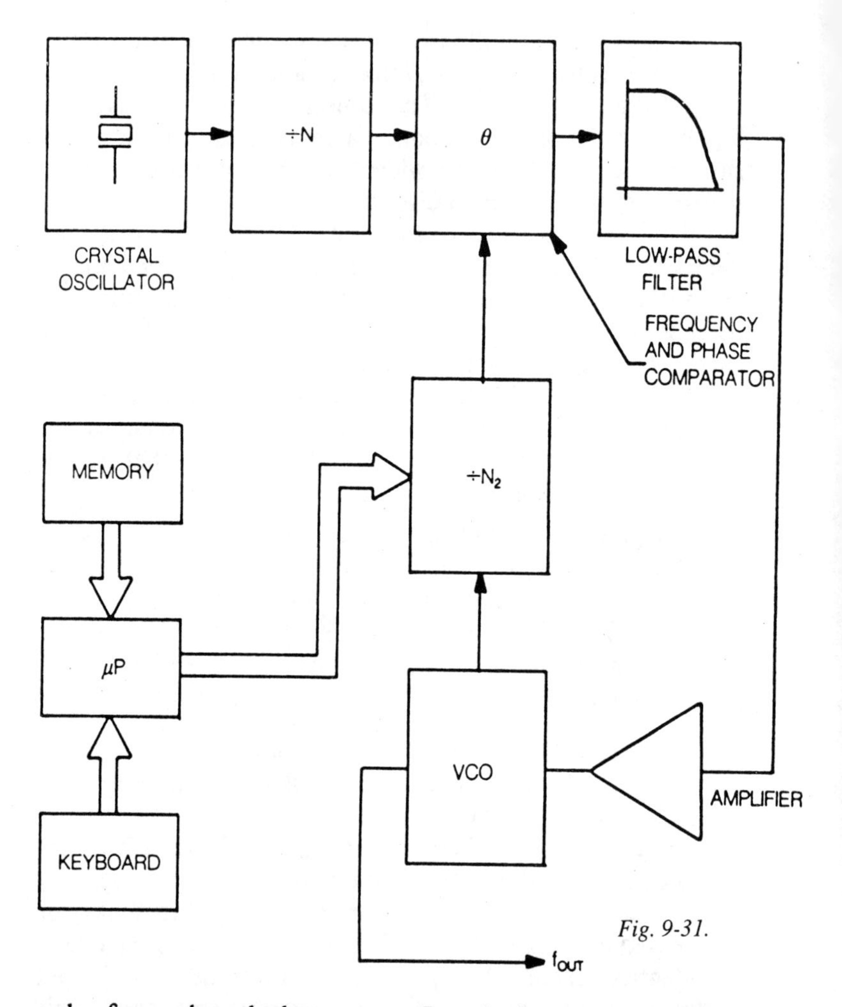

Fig. 9-31.

order for a phase lock to occur. But the input to the phase circuit from crystal has been divided by 2. The combined effect of the two blocks then, is to produce an output at the VCO that is three halves multiplied by the crystal frequency. So, it is possible to produce an output at the VCO that is a fraction or some fractional multiple of the crystal frequency.

You will remember that a microprocessor uses memory. If the codes for the inputs to N2 are put into memory, the microprocessor can be made to deliver to N2 any coded value that has been stored in the memory. The keyboard selects the desired

VCO frequency. The microprocessor gets the proper code from memory and delivers it to the block marked "divide-by-N2." This produces an automatic division to get the desired output frequency.

It is also possible to have the microprocessor control the divide-by-N1 block. In that case, the codes for that divide-by circuit must also be stored in memory. The overall effect is that many output frequencies can be synthesized from the crystal input frequency. The output frequency is highly stable because it depends upon the input crystal frequency.

Frequency synthesizers are used in many applications. Obviously, they are used where a variety of frequencies with crystal accuracy is needed.

An output frequency can be used to control the speed of a motor. Some motors have a speed that is directly related to the frequency of the input power. The output of the VCO must be power amplified to operate those motors.

Two examples of frequency-dependent motors are: synchronous and stepping motors. The synchronous motor was discussed in a previous chapter. Stepping motors rotate a fixed number of degrees with each input pulse.

SUMMARY

Industrial electronics is a traditional field based in part on formal network theorems and laws. When you read magazine articles and company manuals, you will encounter these theorems and laws. They are usually alluded to with the understanding that you know their definitions and terminology.

Some network theorems and laws have been included in this chapter. This is partly a review of material covered earlier with additional emphasis on applications.

An example of applications is in four-terminal network study. The reason for using a four-terminal π or T network is to match input and output impedances for maximum power transfer. This permits the generator side to see its internal resistance, and it allows the output to see the load resistance looking back into the equivalent circuit.

Four-terminal active networks are used extensively in circuit design. Because of this, the parameters often show up in technician's literature. A specific example is *h* parameters, used for defining transistors.

In analyzing networks, Thevenin's and Norton's generators can be very useful tools. The mathematics of the procedure is not of interest here, but we are certainly interested in the ramifications of the procedure. They have been covered in this chapter.

One of the most important concepts in electronics is that of time constants for both RC and RL networks. The equations for calculating time constants have been included in this chapter for reference. Modern calculators make it a relatively easy matter to solve these equations. Some review of time constant circuitry such as used in the UJT has been included.

Integrators and differentiators originally were used in analog computers. Op amp versions of these circuits are especially important for modifying waveforms. These subjects have been reviewed, along with an explanation of the mathematical procedure for integration and differentiation.

No matter which field of electronics you are working in, you are very likely to encounter timers. The 555 timer is a benchmark for integrated circuit timers. This subject was briefly reviewed in this chapter. It was not covered in depth because it was assumed that you already know how they work and you know some basic applications. If you are not familiar with this timer you should review its operation.

While on the subject of benchmarks, the 741 integrated circuit operational amplifier is the benchmark for all op amps. As with the 555, many improvements have been made and new IC's have been introduced. However, they are better understood by comparison with the benchmarks.

In an earlier chapter you studied the principle of frequency synthesizing. The phase-locked loop version has been reviewed and extended in this chapter. With the version given here you can obtain any frequency within a given range. Those frequencies are crystal controlled.

SELF TEST

1. The letter 'h' in h_{FE} stands for:
   (A) high frequency.
   (B) hybrid.

2. To obtain maximum power transfer from an ac generator that has an internal (complex) impedance, the load impedance must be:
   (A) a replica of the internal impedance.
   (B) a conjugate of the internal impedance.

3. The reason you cannot use Ohm's law to solve for the current in a light bulb circuit is that the light bulb resistance is:

   ______________.

4. If you know the phase angle, the power factor can be determined by the equation:

   PF =

5. In a four-terminal passive network there are:
   (A) only four parameters.
   (B) an unlimited number of parameters.

6. Any passive four-terminal network can be replaced with:
   (A) an equivalent T circuit.
   (b) an equivalent π circuit.
   (C) Both choices are correct.
   (D) Neither choice is correct

7. Filter circuits are often named for the procedure used in their design, or for ______________.

8. In the nonmathematical approach for finding a Thevenin equivalent circuit, V_{th} is measured by a:
   (A) function generator.
   (B) high-impedance voltmeter.
   (C) low-impedance voltmeter.
   (D) None of these choices is correct.

9. When a capacitor is being charged through a resistor, it is presumed to be fully charged at the end of ______________ time constants.

10. In a series RL circuit, the time it takes for the current to reach maximum value will be increased by:
    (A) making the resistance of R lower.
    (B) making the resistance of R higher.

ANSWERS TO SELF TEST

1. (B) Hybrid parameters are combinations of voltage and current parameters.

2. (B) A complex impedance is one that has both resistance and reactance.

3. nonlinear.

4. Cos ϕ. Power factor is a measure of how closely voltage and current of a circuit are in phase. The ideal power factor is 1.0. This means that voltage and current are in phase.

5. (A)

6. (C)

7. The name of the originator. An example is Butterworth.

8. (B) The reason a high-impedance meter is needed is that the circuit would be loaded by one having a low impedance. Loading the circuit would result in the wrong voltage reading.

9. five

10. (A) Remember that $T = L/R$. Making the denominator of a fraction a lower value increases the value of the fraction.

10

Measurements and Troubleshooting

THE SUBJECTS OF measurements and troubleshooting are not easily separated. Assuming you have looked for obvious visible faults, troubleshooting starts by making a measurement. It can be a very simple measurement, like judging the quality of a video picture on an intrusion alarm screen, or, it can be a very complicated measurement used for analyzing a system.

After each measurement has been made, you make a judgement. Does the measured value indicate that there is a problem? If so, you might need to make more measurements to narrow the source of the trouble, or your measured value might lead directly to the cause. Once you have located the trouble, your last step is to make the necessary repair.

Before you get to the point where you are studying a specialty like industrial electronics, you have studied general electronics subjects. It is assumed that you know how to use a voltmeter and other basic test equipment. This chapter emphasizes types of measurements and equipment not typical of other fields of electronics. However, some basic techniques that have not been covered in other chapters will also be included.

The Programmed Review in this chapter covers questions that have been frequently missed on CET tests. They are things you are expected to know, but which have not been greatly emphasized throughout the book.

CHAPTER OBJECTIVES

After reviewing the material in this chapter you will be able to answer these questions:

- What are some of the important types of bridges and what are they used for?
- How is a potentiometer used for measuring voltage?
- How much current can a standard cell deliver?
- What is a simple way to use a signal generator (or function generator), a variable resistor, and two voltmeters to measure capacitance?
- What is a Hall device used for?

BRIDGES

You will remember that passive transducers are often connected into bridge circuits. That way, they can be used for sensing without being affected by changes in power supply voltage, or changes in ambient temperature.

When two transducers are used in opposite legs of a bridge, the heating effect of supply current can be cancelled. Also, the measurement can be referenced to the ambient temperature where the sensing takes place. Bridges are also used for making accurate measurements. The Wheatstone bridge is the simplest example.

The Wheatstone Bridge. A *Wheatstone bridge* is an instrument used for accurately measuring resistance. It is shown in its traditional form in Fig. 10-1. The unknown resistance value is R_X. R_1, R_2, and R_3 are precision resistors.

The bridge circuit has been redrawn in Fig. 10-2. The voltage (V_a) at point *a* can be determined by the proportional method as follows:

$$V_a = [R_3/(R_1 + R_3)]V$$

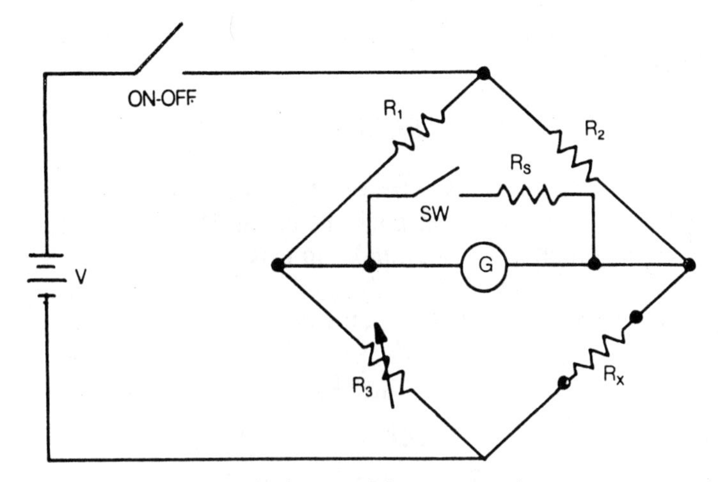

Fig. 10-1. The Wheatstone bridge is used for accurate resistance measurements.

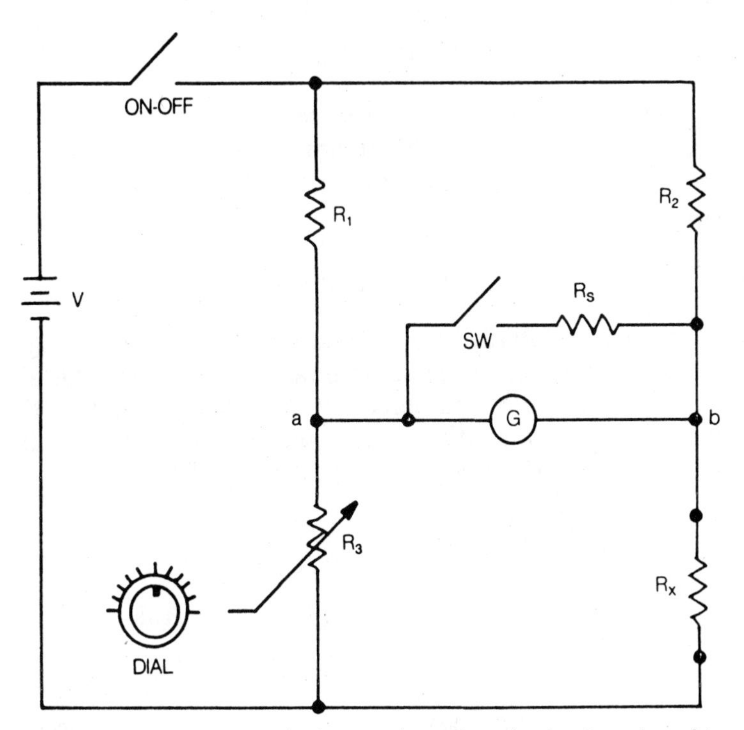

Fig. 10-2. The Wheatstone bridge is redrawn here for the discussion of its operation.

The voltage at point *b* can be determined by the same method:

$$V_b = [R_x/(R_2 + R_x)]V$$

When the bridge is balanced, $V_a = V_b$. Under that condition there is no current through the meter. Resistor R_3 can be adjusted so that a balance condition exists.

A little simple math can be used to express the resistance values when a balance condition exists:

$$V_a = V_b$$

$$[R_3/(R_1 + R_3)]V = [R_x/(R_2 + R_x)]V$$

$$R_x(R_1 + R_3) = R_3(R_2 + R_x)$$

$$R_1R_x = R_2R_3$$

$$R_1/R_2 = R_3/R_x$$

This is an equation for balance. Variable resistor R_3 is adjusted until there is no current through the meter, as required for balance. Then, R_x can be determined from the following equation:

$$R_x = R_2R_3/R_1$$

SAMPLE PROBLEM: Refer to the bridge circuit in Fig. 10-1. When the unknown resistance value—that is, the value of R_x—is connected into the circuit, R_3 must be adjusted to 23K in order to get a reading of zero on the microammeter. If R_1 = 16K, and R_2 = 14K, what is the value of R_x?

SOLUTION:

$$R_x = R_2R_3/R_1$$
$$= (14 \times 23)(K)^2/16K$$
$$= 20.125K$$

The accuracy of measurement depends upon how accurately resistance R_3 can be determined. In professional versions of the bridge, R_3 may be a precision ten-turn variable resistor. The meter movement, marked *G* for galvanometer, may be a fifty microampere (full-scale) instrument. The fixed resistors R_1 and R_2 will have precision resistance values.

The equation for balance can be rewritten as follows:

$$R_3 = R_1 R_X / R_2$$

The fixed resistors (R_1/R_2) can be laser trimmed to ensure exact resistance values. That, in turn, helps to ensure the accuracy of bridge measurements. The bypass resistor R_5 around the meter movement serves as a meter shunt. It converts the sensitive meter to a miliammeter.

Assume that a resistor with an unknown resistance value is connected into the position for R_x. This is usually done with thumbscrew terminals that make a very good electrical connection.

When the on-off switch (SW1) is first operated, the bridge may be very far off balance. This would cause the sensitive meter movement to be destroyed if switch SW2 is open. If SW2 is closed, the meter movement is shunted. Then, its lower sensitivity allows the meter to deflect, but not enough to damage the meter movement.

Resistor R_3 is adjusted with SW2 closed so that a near balance condition exists. Then, SW2 is opened to increase the sensitivity of the current measurement. A final adjustment of R_3 is made to get the most sensitive measurement.

Other Bridges. The Wheatstone bridge cannot be used to measure very low resistance values. The reason is that the terminals on the bridge combine with the low value of resistance being measured. This makes it impossible to tell which is contact resistance and which is the resistance being measured.

A special version of the Wheatstone bridge, called the Kelvin bridge, makes it possible to balance out the low value of contact resistance. It can be used for accurately measuring very low resistance values.

AC Bridges. To measure inductance or capacitance by the bridge method, it is necessary to use an ac voltage source instead of the battery that is used with a Wheatstone bridge. Many forms of L and C bridges are available.

The Maxwell-Wein bridge of Fig. 10-3 is designed for measuring inductance. It is also known as a Maxwell bridge.

Since an inductor, in practice, can always be represented by an inductor in series with a resistor, as shown in the lower right-hand side of the Maxwell bridge, the bridge must be balanced for both L and R. That results in two equations for the balance condition:

$$L = (R_2R_4)$$

and

$$R = R_2R_3/R_1$$

Resistors R_2, R_3, and R_4 are precision noninductive types.

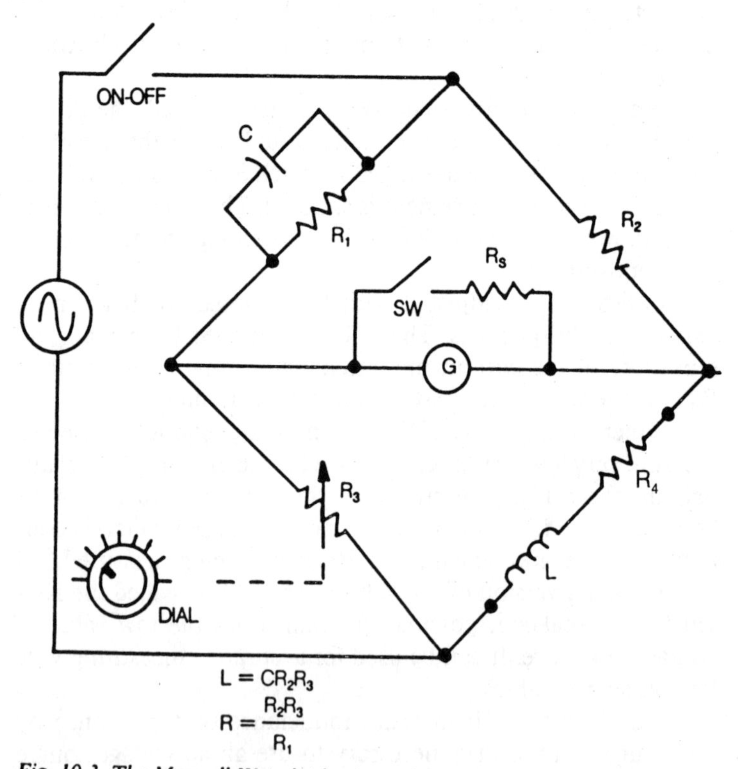

Fig. 10-3. The Maxwell-Wein bridge is used for measuring inductance.

Balancing the bridge is not an easy task if it is being accomplished by adjusting the resistors. However, if *C* is variable along with R_1, balance is easily obtained.

You may wonder why you should bother with an inductance bridge when digital LCR meters are readily available. The answer is that you need to know both the L and the R of an inductor in order to get a true picture of it. Also, commercial bridges may also give the Q (X_L/R) and other parameters not available with the LCR meter.

If the Q of the inductor is high—that is, if X_L is a high value with respect to R—then a modification of the Maxwell-Wein bridge is preferred. It is the Hay bridge shown in Fig. 10-4.

As shown in the illustration, it is necessary to know the frequency in order to get a precise value of inductance and resistance. However, a good approximation can be obtained by

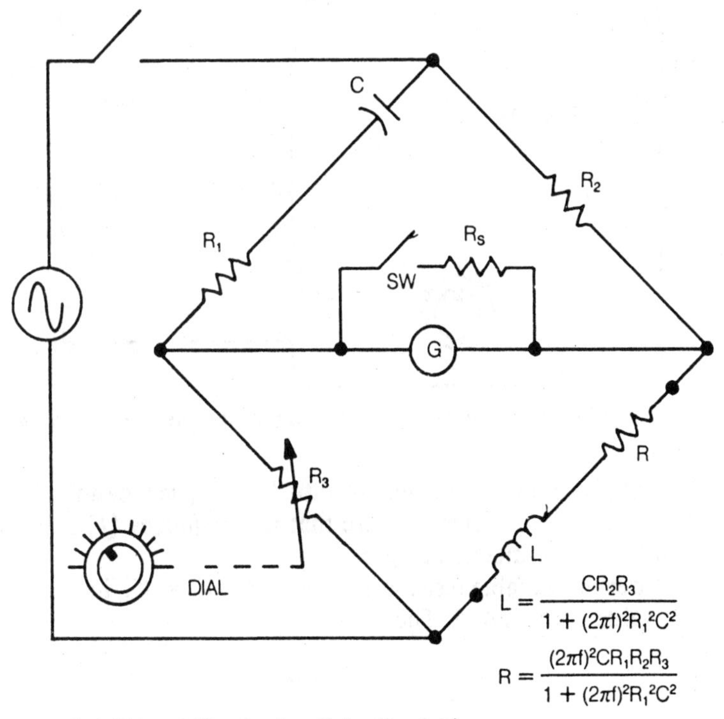

Fig. 10-4. This modification is called a Hay bridge.

disregarding the 2Π term. If the values of L and R are accurately known, then C and R_l can be obtained by rearranging the bridge equations.

Voltage-Resistance Measurement of *L* or *C*. You can get a reasonably accurate measurement of inductance or capacitance by using the test setup in Fig. 10-5. You need to use a frequency that will give a reasonably high value of X, which is the inductive or capacitive reactance of the unknown.

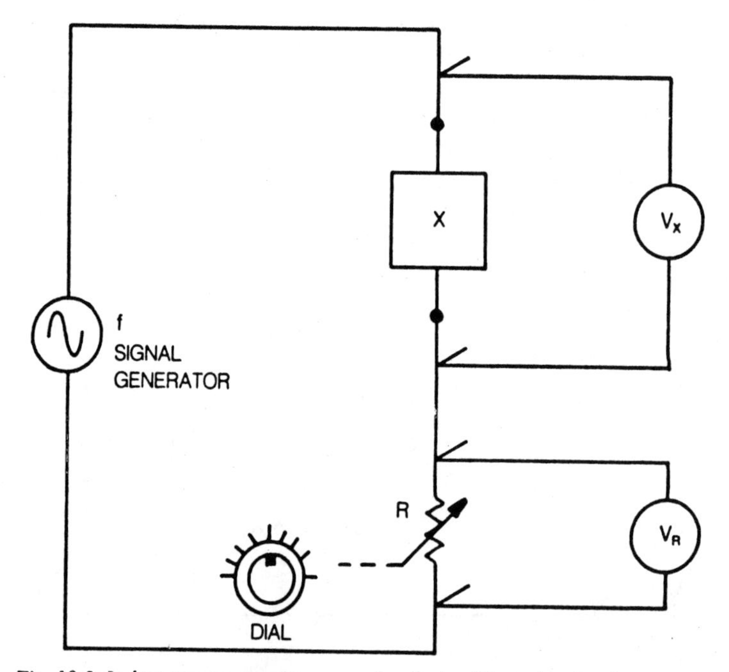

Fig. 10-5. Inductance or capacitance can be obtained from this simple test setup.

The procedure is to adjust R until $V_x = V_R$ as measured by a high-impedance meter. Be sure that the frequency (f) is not too high for the meters being used.

After the variable resistor is adjusted for $V_x = V_R$, its resistance value is measured. Then,

$$X_L = R$$

or,

$$2\Pi fL = R$$
$$L = \mathrm{R}/2\Pi f$$

Also,

$$X_C = R$$

or,

$$1/2\Pi fC = R$$
$$C = 1/2\Pi fR$$

EXACT VOLTAGE MEASUREMENTS

In commerce, it is necessary to have exact, reliable measurements. That permits each country to accurately determine price per unit of goods.

In the past in France, each seaport had its own standard of measurement. It was very difficult—if not impossible—to convert each seaport's units of measurement to those of a trader. One very important reason for developing the metric system was to eliminate that confusion.

Bringing this up to date, if countries are trading in electronic systems, they must have electrical standards. We could not afford the confusion that would exist by having a different meaning for the volt in every country.

For that reason, each country has the equivalent of a Bureau of Standards. Matching the standards of various countries is an on-going task.

Large companies often have a secondary standards lab that is matched with the Bureau's primary standard. This is especially true of large companies that make electronic equipment for the U.S. military.

Calibration of test equipment is one of the functions of a secondary standards lab. Contracts usually specify how often that calibration is required. The secondary standards lab keeps very precise components such as resistors, capacitors, and inductors. They are used for calibration procedures.

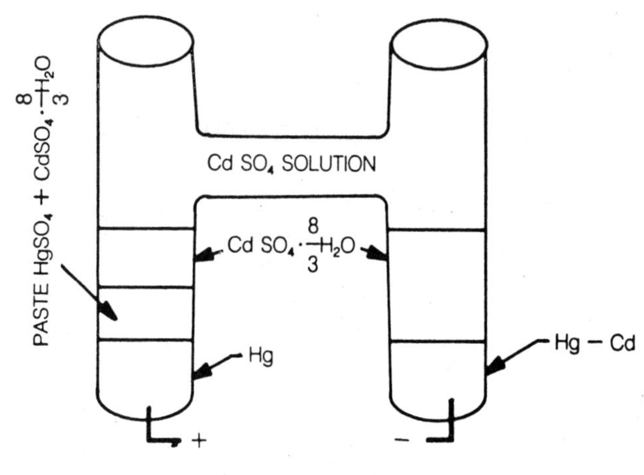

Fig. 10-6. The standard cell is used as a secondary standard for voltage around the world.

The Standard Cell. Figure 10-6 shows a standard cell. It has an exact voltage value, so it can be used for precise measurement and calibration.

The first thing you should know and remember about a standard cell is that you must not touch the platinum leads! Also these cells must never be put into a circuit that draws more than 0.0001 ampere. If you try to measure the cell's voltage with a voltmeter you will destroy it!

One method of using a standard cell is to allow it to very slowly charge a capacitor. The voltage across the capacitor is then used for a voltage standard.

Another method of using the standard cell is to employ it as a reference in an instrument called a potentiometer. You have used that term to refer to variable resistors, but the voltage-measuring instrument is not related to that component.

A SIMPLE POTENTIOMETER

Figure 10-7 shows a simple potentiometer circuit. It consists of two loops: ABCDA and EFGCE. Loop ABCDA provides a fixed dc current through the resistor. This is important to remember about this circuit.

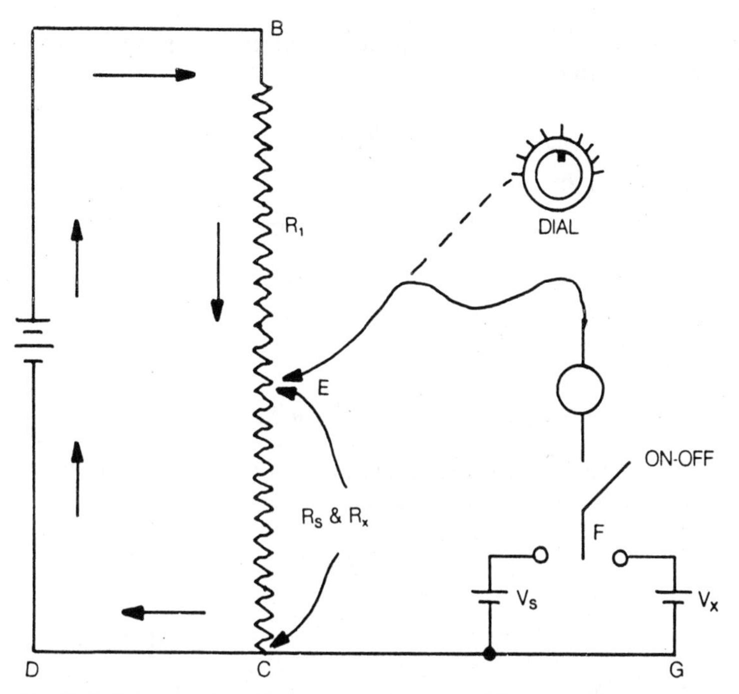

Fig. 10-7. Potentiometers, like this example, are used for accurately measuring an unknown voltage.

Two sources of voltage are in the second loop. Voltage source V_s is the standard cell. Its voltage is accurate known. Voltage source V_x is the unknown voltage source. Its voltage value is to be measured. The position of the tap on the resistor is accurately known and is displayed by the dial.

When the two-way switch at F is in the left position, the voltage of the standard cell is matched against the voltage at e. The variable resistor is adjusted so that no current flows in the galvanometer. The resistance required to produce zero current is noted and it will be called R_S from here on.

Next, the switch at position F is set to the right side. Now V_x is in the circuit and matched against the voltage at e. The variable resistor is adjusted again until there is no current indicated by the galvanometer. This resistance setting is noted on the dial and called R_x.

Using the fixed value of conventional current flow, the voltage at point e for the setting of R_S is IR_S. The only voltage at the setting of R_x is IR_x. Since no current flows for those settings, it follows that:

$$V_s = IR_s$$

Solving for I:

$$I = V_s/R_s$$

Also,

$$V_x = IR_x$$

and

$$I = V_x/R_x$$

The value of I is the same during both adjustments. Therefore, the equations can be solved as follows:

$$I = I$$

so:

$$V_s/R_s = V_x/R_x$$

Solving for V_x:

$$V_x = V_s R_x/R_s$$

All the terms of the right side of the equation are accurately known. So, the unknown voltage (V_x) can be accurately determined.

There is an additional feature of the potentiometer not shown in Fig. 10-7. You will remember that the standard cell must not deliver a current greater than 0.0001 amperes.

Some provision must be made to protect the standard cell against a greater current when the circuit is first energized. Here are some ways of doing that:

- One way is to connect a large resistance value in series with the cell to limit current to a very low value until an initial balance can be obtained.
- A rheostat can be used in series with the signal generator to limit cell current until an initial balance can be obtained.

- A more rugged source of voltage can be substituted for V_s until an initial balance can be obtained.

Regardless of the method used to protect V_S, the basic principle of the potentiometer is the same as described in this section.

SUMMARY

In this chapter you have reviewed some of the very important methods of making measurements. Specifically, the use of bridges and potentiometers has been covered. They provide very accurate measurements of resistance and voltage. Of course, using R and V, you can find the current, so all of the elements of Ohm's law are available from those measurements. This is the foundation of many of the measurements used in industry.

Bridges are used in other applications besides measurements. In an earlier chapter you reviewed them in transducer circuits. They are also important in making measurements.

You will now go through a Programmed Review. It takes the place of a Self Test for this chapter. It covers some very important subjects that should be emphasized, and some new ideas.

PROGRAMMED REVIEW

Start with block number 1. Pick the answer that you feel is correct. If you select choice number 1, go to block 13. If you select choice number 2, go to block 15. Proceed as directed. There is only one correct answer for each question.

BLOCK 1

Is the following statement correct? You cannot measure current with an oscilloscope.

(A) The statement is not correct. Go to block 15.

(B) The statement is correct. Go to block 13.

BLOCK 2

Your answer to the question in block 8 is not correct. Go back and read the question again and select another answer.

BLOCK 3

Your answer to the question in block 15 is not correct. Go back and read the question again and select another answer.

BLOCK 4

Your answer to the question in block 21 is not correct. Go back and read the question again and select another answer.

BLOCK 5

Your answer to the question in block 11 is not correct. Go back and read the question again and select another answer.

BLOCK 6

The correct answer to the question in block 16 is choice (B). Four-layer diodes are thyristors.

Here is your next question: Which of the field-effect transistors shown in this block requires a reverse bias when used as a Class A amplifier?

(A) The one marked A. Go to block 26.
(B) The one marked B. Go to block 30.

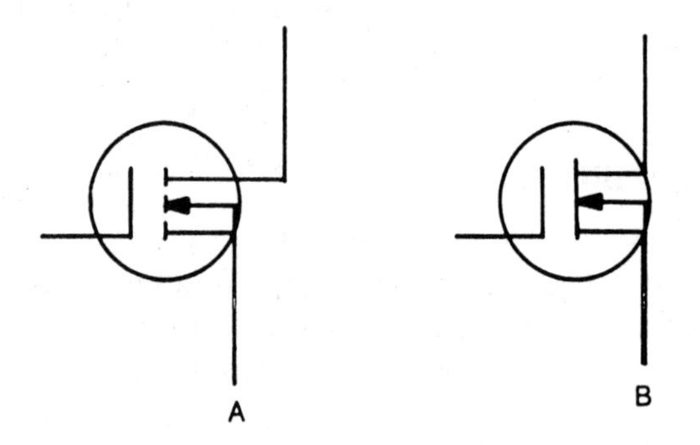

BLOCK 7

Your answer to the question in block 23 is not correct. Go back and read the question again and select another answer.

BLOCK 8

The correct answer to the question in block 18 is choice (B). The sum of the currents (I_E) a junction transistor equals the sum of the currents ($I_B + I_C$) leaving the junction transistor.

Here is your next question: What voltage reading should you get in the test setup shown in this block? The meter resistance is 100K as shown.

(A) 50.8V. Go to block 23.
(B) 31.3V. Go to block 2.
(C) 62.6V. Go to block 17.

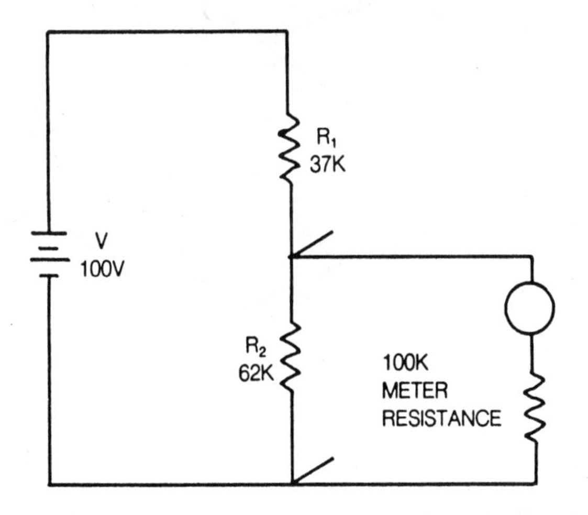

BLOCK 9

Your answer to the question in block 21 is not correct. Go back and read the question again and select another answer.

BLOCK 10

Your answer to the question in block 20 is not correct. Go back and read the question again and select another answer.

BLOCK 11

The correct answer to the question in block 23 is choice (D). The relay coil will produce inductive kickback that will destroy the transistor. The thermistor will not protect the transistor. Replace it with a diode or a varistor.

Here is your next question: Which of the following is a use for the device illustrated in this block?

(A) Sense magnetic field. Go to block 20.
(B) Produce magnetic field. Go to block 5.

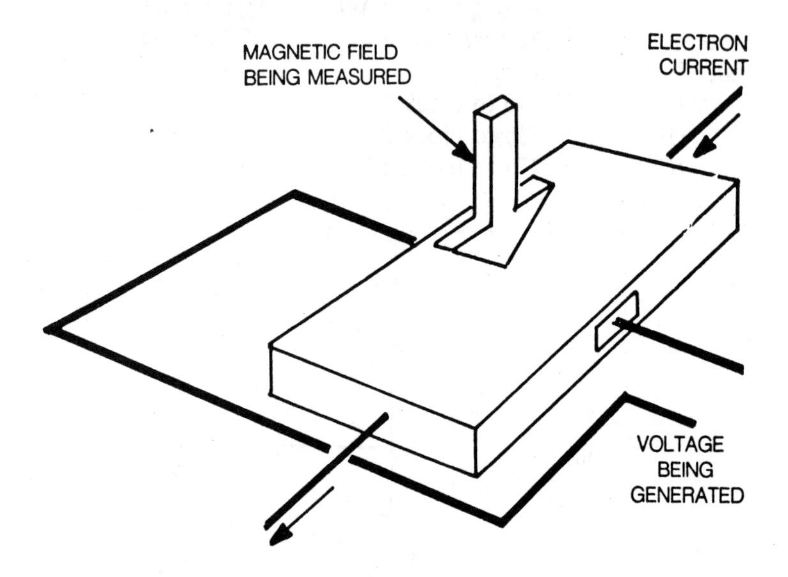

BLOCK 12

Your answer to the question in block 18 is not correct. Go back and read the question again and select another answer.

BLOCK 13

Your answer to the question in block 1 is not correct. Go back and read the question again and select another answer.

BLOCK 14

Your answer to the question in block 16 is not correct. Go back and read the question again and select another answer.

BLOCK 15

The correct answer to the question in block 1 is (A). The illustration in this block shows the procedure. If you cannot get enough deflection on the scope, try a ten-ohm resistor.

Here is your next question: An oscilloscope display of a

sawtooth voltage is shown in this block. The RMS value of the sawtooth is:

(A) 7.07V. Go to block 25.
(B) 3.53V. Go to block 3.
(C) Neither choice is correct. Go to block 21.

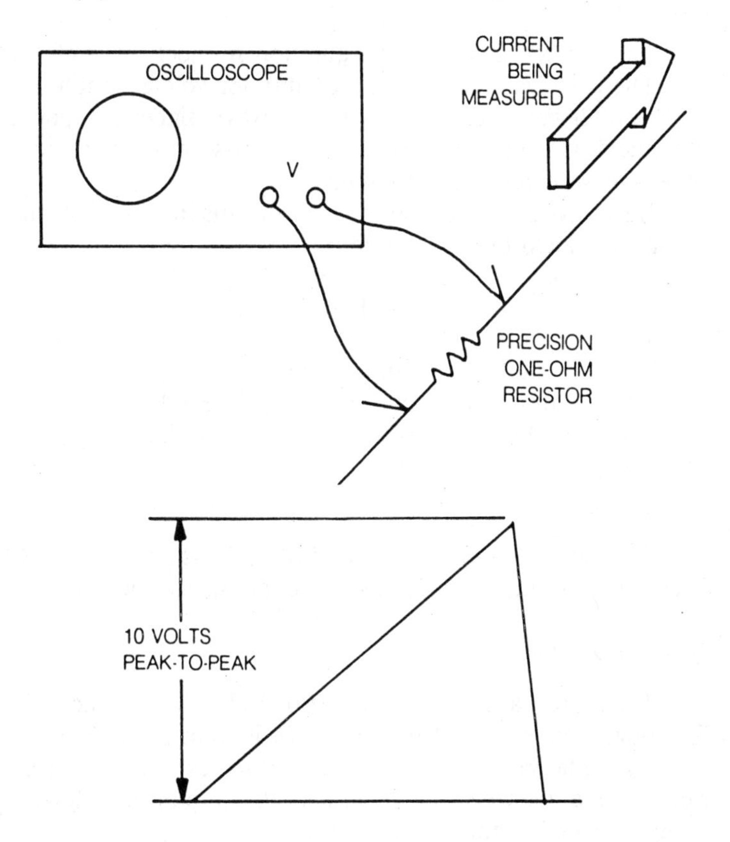

BLOCK 16

The correct answer to the question in block 20 is choice (B). It does not matter what combination of 1's and 0's is delivered to the input terminals, the output will always be a logic 1.

Here is your next question: A Shockley diode is also called:

(A) a three-layer diode. Go to block 14.
(B) a four-layer diode. Go to block 6.

BLOCK 17

Your answer to the question in block 8 is not correct. Read the question again and select another answer.

BLOCK 18

The correct answer to the question in block 21 is choice (C). The 0.7V value is a ballpark figure for voltage amplifiers. The emitter-base voltage is almost always higher for power amplifiers. If your measurement does not seem correct, make other measurements in the circuit.

Here is your next question: The currents in a bipolar transistor are related by the equation:

$$I_e = I_B + I_C$$

This equation is based upon:

(A) Ohm's law for current. Go to block 12.
(B) Kirchhoff's law for current. Go to block 8.

BLOCK 19

Your answer to the question in block 23 is not correct. Go back and read the question and select another answer.

BLOCK 20

The correct answer to the question in block 11 is choice A. The device shown is a Hall device. In its normal operation a current (I) flows through the device. When exposed to a magnetic field, a voltage (V) is generated. the amount of voltage is directly related to the magnetic field.

In operation, a voltmeter across the terminals marked V is calibrated to read the amount of magnetism.

Here is your next question: The logic probe in the circuit for this block should show the presence of a:

(A) logic 0. Go to block 28.
(B) logic 1. Go to block 16.
(C) pulse. Go to block 10.

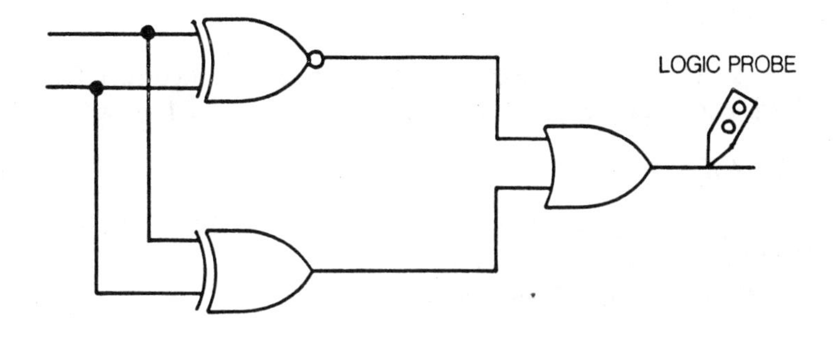

BLOCK 21

The correct answer to the question in block 15 is choice (C). The equations for RMS voltage, shown below, are for pure sine waves only.

Here are the equations for RMS and average values of a pure sine waveform:

$$V = 0.707 \times V_{max}$$
$$V = 0.636 \times V_{max}$$
$$V = 0.707 \times V_{p\text{-}p}/2$$
$$V = 0.636 \times V_{p\text{-}p}/2$$

where: V is the RMS value,
V_{max} is the peak value,
$V_{p\text{-}p}$ is the peak-to-peak value.

Here is your next question: The emitter-to-base voltage is measured on a silicon Class A power amplifier. Its value is 1.1 volts. Which of the following is a correct interpretation of this measurement?

(A) The transistor is defective. Silicon transistors have a 0.7V emitter-base voltage. Go to block 9.

(B) The measurement shows that the transistor is overdriven. Go to block 4.

(C) There is no problem indicated by this measurement. Go to block 18.

(D) None of these choices is correct. Go to block 24.

BLOCK 22

Your answer to the question in block 23 is not correct. Go back and read the question again and select another answer.

BLOCK 23

The correct answer to the question in block 8 is choice (A). Here is one way of solving the problem:

1. Find the circuit resistance (R_T)

$$R_T = 37K + (62 \times 100)/(62 + 100)K = 75.27K$$

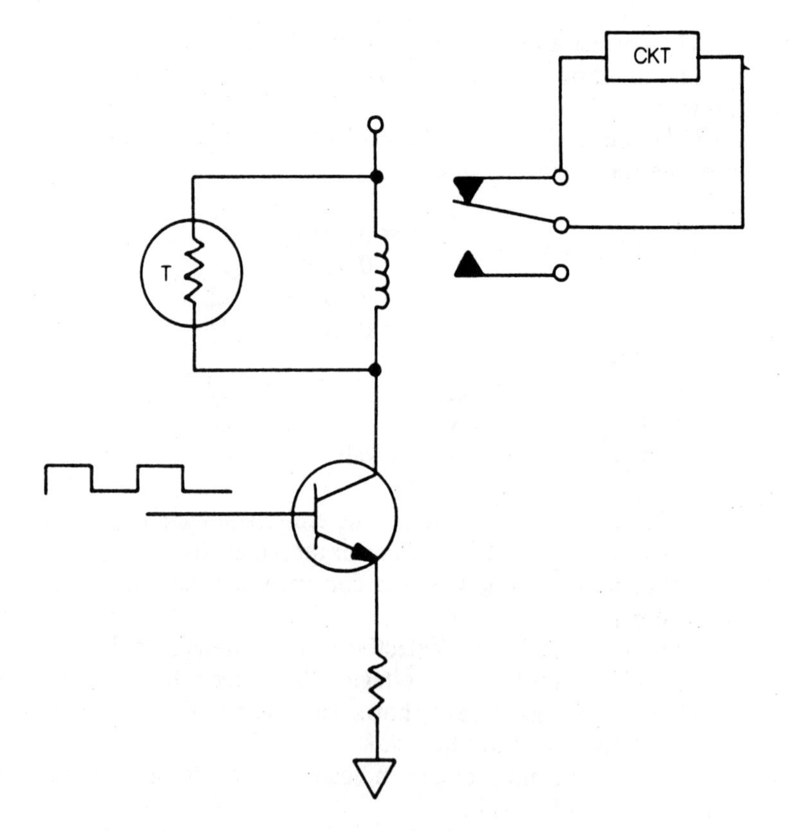

2. Find the current supplied by the battery (I_T).

$$I_T = 100V/R_T = 100/75.27K$$

$= 1.328$ milliamperes

3. Find the voltage drop (V_1) across R_1.

$V_1 = 1.32 \times 10^3 \times 37 \times 10^{-3}$
$V_1 = 49.153$

4. Subtract V_1 from 100 to get the voltage across the meter—50.8V.

Here is your next question: The transistor in the experimental circuit shown in this block falls after a few input pulses. A likely cause is:

(A) poor transistor quality. Go to block 22.
(B) wrong transistor type. Go to block 7.
(C) resistor R does not have the proper resistance. Go to block 19.
(D) None of these choices is correct. Go to block 11.

BLOCK 24

Your answer to the question in block 21 is not correct. Go back and read the question again and select another answer.

BLOCK 25

Your answer to the question in block 15 is not correct. Go back and read the question again and select another answer.

BLOCK 26

Your answer to the question in block 6 is not correct. Go back and read the question again and select another answer.

BLOCK 27

The correct answer to the question in block 32 is choice (A).

An A/D converter converts an analog signal (like a dc voltage) into a digital signal (like a signal that represents a binary number).

Analog-to-digital (A/D) converters and digital-to-analog (D/A) converters are available in integrated circuit packages. They are used in digital systems applied to industrial electronics. You have now completed the Programmed Review.

BLOCK 28

Your answer to the question in block 20 is not correct. Go back and read the question again and select another answer.

BLOCK 29

Your answer to the question in block 30 is not correct. Go back and read the question again and select another answer.

BLOCK 30

The correct answer to the question in block 6 is choice B. In order to be able to troubleshoot effectively, you must know the components and the voltage polarities required for their operation.

Here is your next question: The frequency of a UJT will increase when the resistance in its time constant circuit is:

(A) decreased. Go to block 31.
(B) increased. Go to block 29.

BLOCK 31

The correct answer to the question in block 30 is choice (A). Decreasing the resistance allows the capacitor to charge faster. That, in turn, reduces the time (T) for one cycle. Since $f = 1/T$, reducing T increases f.

Here is your next question: Write the equation for duty cycle.

Duty cycle =

Go to block 32.

BLOCK 32

Here is the correct answer to the question in block 31.

Duty cycle = ON TIME / TIME FOR COMPLETE CYCLE

When multiplied by 100, it is percent duty cycle. Duty cycle is sometimes called duty factor.

Here is your next question: Which of the following would be useful for converting dc to a binary number?

(A) A/D converter. Go to block 27.
(B) D/A converter. Go to block 33.

BLOCK 33

Your answer to the question in block 32 is not correct. Go back and read the question again and select another answer.

Appendix

Practice CET Test:

Industrial Electronics Option

THE CET TEST that you take may be divided into the following sections:

- Components for Industrial Electronics,
- Analog Circuits and Systems,
- Digital and Microprocessor Circuits and Systems,
- Dc Power Supplies,
- Ac Power Supplies, and
- Troubleshooting and Circuit Analysis.

These subjects have been reviewed in this book. Also, they will be covered in this Practice Test. However, this test is not divided into those sections. The number of questions on each subject is related to the difficulty and number of times questions on the subjects have been missed rather than on sections in the test.

As with all practice tests, this one covers the range of subjects. These are not the questions you will actually get when you take the test.

1. The circuit in Fig. A-1 can be used as:

   (A) A low-pass filter.
   (B) An integrator.
   (C) A differentiator.
   (D) None of these choices is correct.

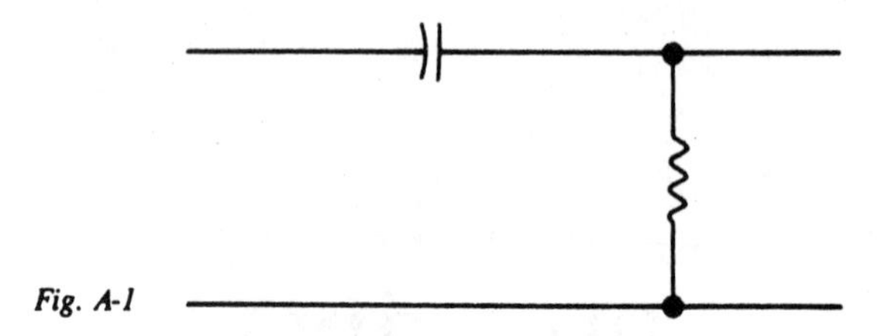

Fig. A-1

2. Which of the following is not correct?

   (A) I = V/R.
   (B) The parallel resistance of two resistors in parallel = $\frac{1}{\frac{1}{R1}+\frac{1}{R2}}$.
   (C) f = 1/T.
   (D) Transformer turns ratio is equal to N_S/N_P.

3. In a VCO circuit you would expect to find:

   (A) A varistor.
   (B) An avalanche diode.
   (C) A Shockley diode.
   (D) A varactor.

4. The power supply regulator in Fig. A-2 is a/an:

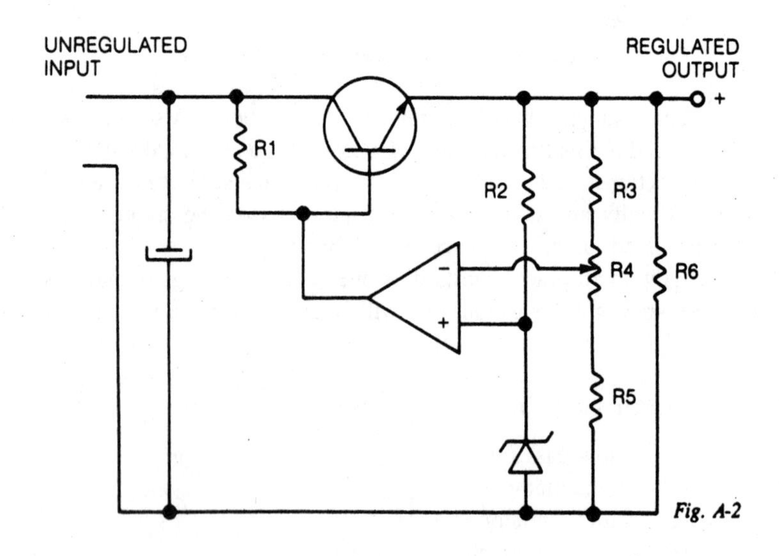

Fig. A-2

(A) Analog type.
(B) Digital type.
(C) Positive feedback type.
(D) Differential type.

5. For the power supply circuit in Fig. A-2, the sense circuit is:

(A) R_1.
(B) R_2 and D_1.
(C) R_3, R_4, and R_5.
(D) R_6.

6. The rate of change of output signal voltage for a step-voltage input is called:

(A) Differential gain.
(B) Integral output.
(C) Acceleration recovery.
(D) Slewing rate.

7. In the circuit of Fig. A-2, the + and − signs on the op amp refer to:

(A) The polarities of the inputs.
(B) The connection of the op amp for inverting or noninverting operation.
(C) Both choices are correct.

8. In a certain system a logic 1 delivered to the input of the flip flop results in a logic 1 at Q. When a logic 0 is delivered to the input of the same flip flop there is a logic 0 at Q.

What type of flip flop is in this system?

(A) R-S.
(B) Eccles Jordan.
(C) Reed Lowey.
(D) None of these choices is correct.

9. The transistor arrangement in Fig. A-3 is often made with two matched transistors in the same case. The difficulty with this arrangement is:

(A) Poor power gain.

(B) False triggering.
(C) Very high input power requirement.
(D) High internal heat.

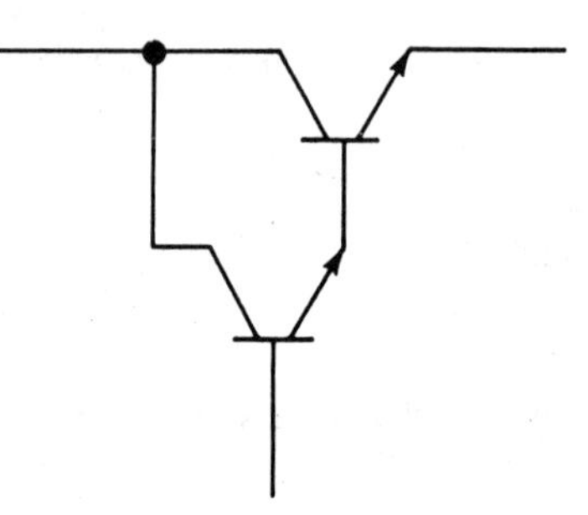

Fig. A-3

10. The transistor arrangement in Fig. A-3 is used as a power supply series-pass regulator. It is called:

(A) Daytona.
(B) Darlington.
(C) Complementary.
(D) Operational.

11. Refer again to the transistor arrangement in Fig. A-3; this arrangement can also be called:

(A) Divide-by-beta.
(B) Parallel-connected.
(C) Beta squared.
(D) Beta + beta.

12. Refer to the circuits shown in Fig. A-4. Assume the power amplifiers are working properly. Which diode connection protects the transistor from inductive kickback?

(A) The one shown in A.
(B) The one shown in B.
(C) Neither choice is correct because the diodes have nothing to do with inductive kickback.

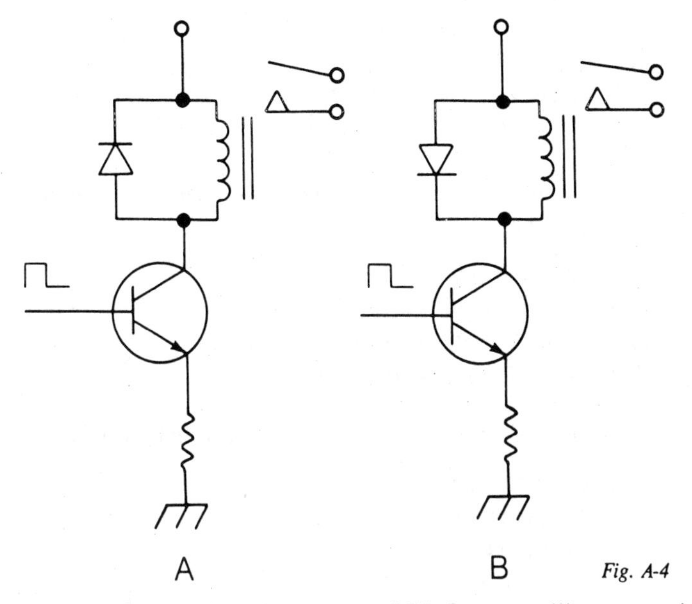

Fig. A-4

13. An advantage of a relay over an SCR for controlling power is that:

(A) It is faster.
(B) It has a higher fan-out capability.

14. The output of the PLL in Fig. A-5 is:

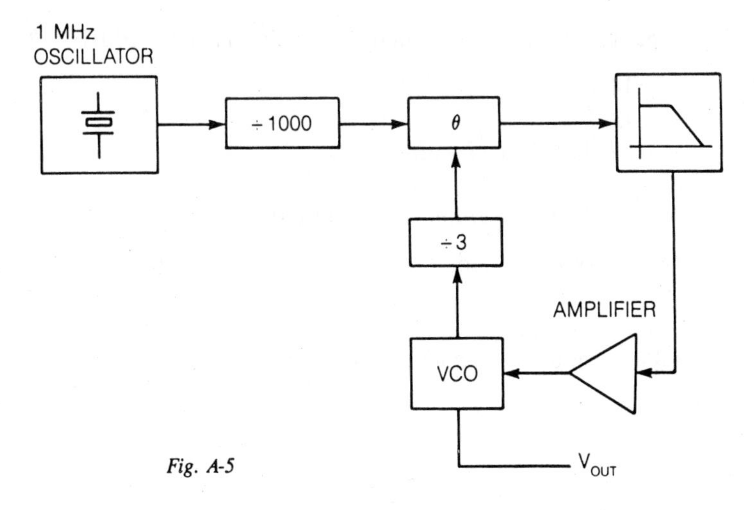

Fig. A-5

(A) 1000 Hz.
(B) 3 MHz.
(C) 3000 Hz.
(D) 333-1/3 Hz.

15. Which of the following will reverse the direction of rotation of an induction motor?

(A) Reverse the armature leads.
(B) Reverse the field leads.
(C) Reverse the armature and field leads.
(D) None of the choices is correct.

16. You would expect to find a rotating magnetic field inside:

(A) A capacitor-start ac motor.
(B) An induction motor.
(C) Both choices are correct.
(D) Neither choice is correct.

17. Power amplifiers:

(A) Are not as much affected by power supply ripple as voltage amplifiers.
(B) Are more affected by power supply ripple than voltage amplifiers.

18. What number is represented by binary decimal code 10010011?

(A) 147.
(B) 43.
(C) Neither choice is correct.

19. You would expect to find a crowbar circuit in a:

(A) Phase-locked loop.
(B) DMA circuit.
(C) Power supply.
(D) Dynamic random access memory.

20. Which of the following is represented by the truth table in Fig. A-6?

(A) $\overline{A} + \overline{B} = L$.
(B) $\overline{A + B} = L$.
(C) $A\,B = L$.
(D) $A + B = L$.

A	B	L
0	0	1
0	1	0
1	0	0
1	1	0

Fig. A-6

21. Which of the following gates is represented by the truth table in Fig. A-6?

(A) AND.
(B) OR.
(C) NAND.
(D) NOR.

22. In this discussion the turns ratio of a transformer is the number of secondary turns divided by the number of primary turns. This is often written $N_s:N_p$.

Which of the following is correct for a power transformer?

(A) $N_s:N_p = I_s:I_p$.
(B) $N_s:N_p = Z_s:Z_p$.
(C) $N_s:N_p = V_p:V_s$.
(D) None of these choices is correct.

(Note: Make sure you know the relationships between turns ratio and current ratio, voltage ratio, and impedance ratio.)

23. Which of the following gates is represented by the circuit in Fig. A-7?

(A) Inverter.
(B) NOR.
(C) EXCLUSIVE OR.
(D) None of these choices is correct.

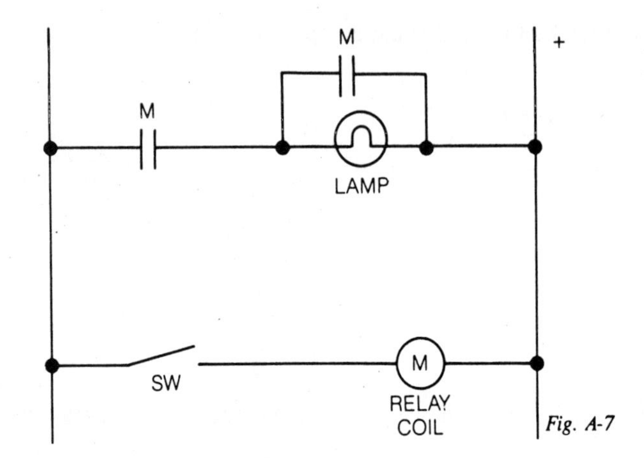

Fig. A-7

24. Refer to Fig. A-7.

    (A) The lamp will be ON when the switch is closed.
    (B) The lamp is ON when the switch is opened.
    (C) Neither choice is correct.

25. A tachometer measures:

    (A) Torque.
    (B) RPM.
    (C) Linear speed.
    (D) Input power.

26. Refer to the illustration in Fig. A-8. The output signal at V_{out} should be a:

    (A) Sawtooth waveform.
    (B) Positive pulse.
    (C) Negative pulse.
    (D) Square wave.

27. The intrinsic standoff ratio of the UJT in Fig. A-8 is 0.63. The frequency of oscillation should be about:

    (A) 1000 Hz.
    (B) 2000 Hz.
    (C) 3000 Hz.
    (D) 5000 Hz.

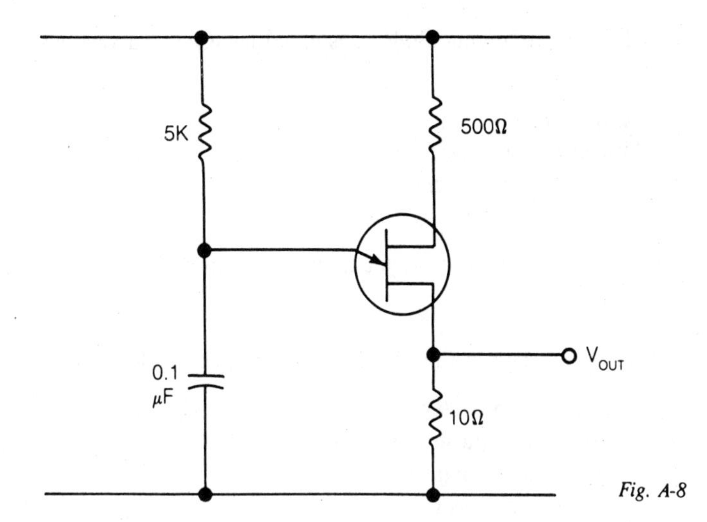

Fig. A-8

28. The symbol shown in the box in Fig. A-9 represents the same device as the one shown in:

(A) A.
(B) B.
(C) C.
(D) D.

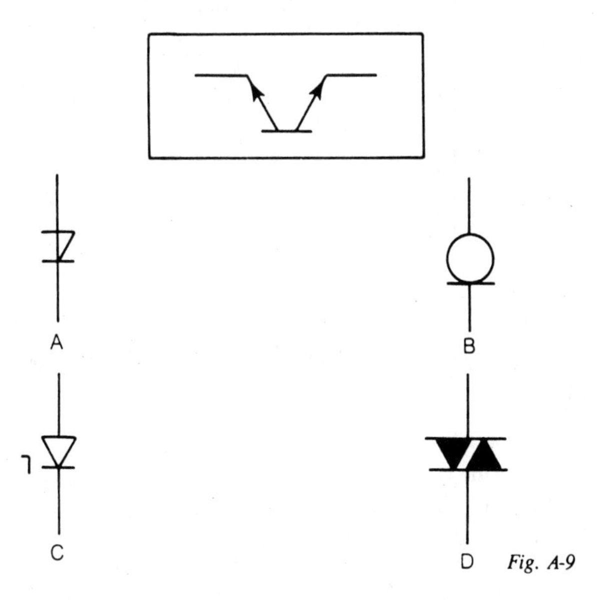

Fig. A-9

29. Which of the following diodes is sometimes used as a very fast switch?

(A) Esaki (tunnel) diode.
(B) Shockley (four-layer) diode.
(C) Schottky (hot carrier) diode.
(D) Point contact diode.

(Important note: Make sure you know about all of the choices in this question.)

30. Refer to the circuit in Fig. A-10. The output (V_{out}) should be a:

(A) Pure dc at a predetermined level.
(B) Positive-going pulse waveform.
(C) Negative-going pulse waveform.
(D) Sawtooth waveform.

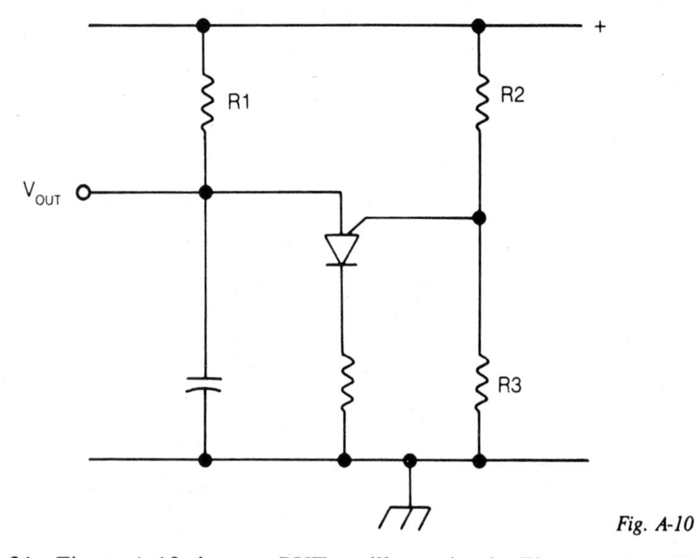

Fig. A-10

31. Figure A-10 shows a PUT oscillator circuit. The purpose of R_2 and R_3 is:

(A) To lower the anode voltage.
(B) To set the dc output voltage.
(C) To set the intrinsic standoff ratio.
(D) None of the choices is correct.

32. Refer to the circuit of Fig. A-11. When the arm of R is moved all the way to X:

(A) The SCR will be destroyed.
(B) The lamp will glow at full brightness.
(C) The lamp will glow at half brightness.
(D) The lamp will be OFF.

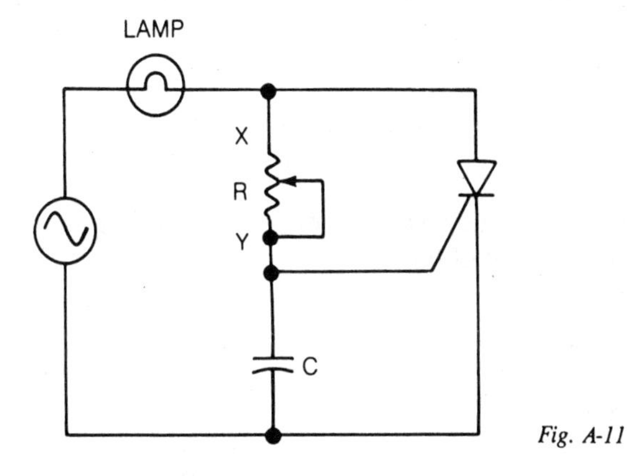

Fig. A-11

33. Which of the following is true about the symbol in Fig. A-12?

(A) $\overline{A}\,B + A\,\overline{B} = L$.
(B) $\overline{A\,B} + A\,B = L$.
(C) $\overline{A + B} = L$.
(D) $\overline{A} + \overline{B} = L$.

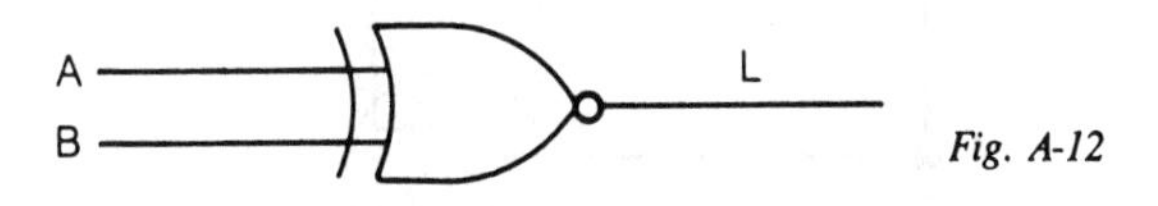

Fig. A-12

34. Is the following statement correct? A Hall device can be used to measure the strength of a magnetic field.

(A) Correct.
(B) Not correct.

35. Form A contacts on a relay (or other switch) are:

(A) Normally open.
(B) Normally closed.

36. You cannot write information into:

(A) An EPROM.
(B) A ROM.
(C) An EEROM.
(D) A RAM.

37. Which of the following is the conjugate of 23 - j17?

(A) 23.
(B) j17.
(C) 23 - j17.
(D) 23 + j17.

38. What approximate voltage reading would you expect to get at Point A in the circuit of Fig. A-13?

(A) 0.7 V.
(B) 1.7 V.
(C) 2.7 V.
(D) 3.7 V.

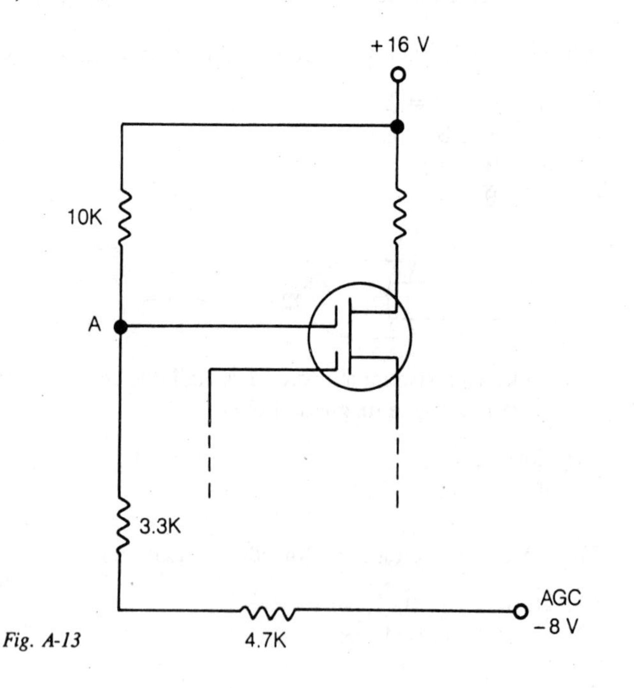

Fig. A-13

39. A 1500-Hz square wave is applied to the input of an amplifier circuit. The output signal is displayed on an oscilloscope and it looks like the waveform in Fig. A-14. The amplifier has:

(A) Poor low-frequency response.
(B) Poor high-frequency response.

Fig. A-14

40. To troubleshoot a closed-loop system:

(A) Short circuit the output and look for excessive power dissipation.
(B) Provide a step change in the input and look for oscillation.
(C) Provide a step change in the output impedance and look for oscillation.
(D) Open the loop and provide the proper substitute voltage or frequency.

41. In the simple arrangement shown in Fig. A-15 the dc motor speed can be varied by:

(A) Changing the pulse frequency.
(B) Changing the pulse width.

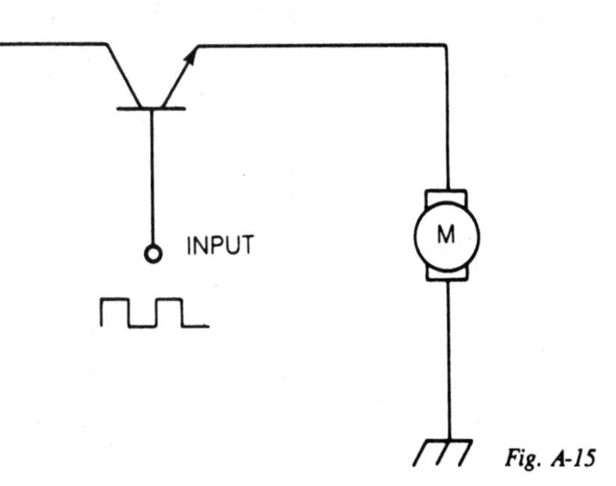

Fig. A-15

42. Brightness on an oscilloscope can be controlled by the proper input to the:

(A) X-axis.
(B) Y-axis.
(C) Z-axis.
(D) None of these choices is correct.

43. A rotary converter is used for:

(A) Increasing ac voltage.
(B) Increasing ac power.
(C) Increasing dc voltage.
(D) Correcting phase angle.

44. Which of the following is true?

(A) Power factor must be greater than 1.0.
(B) Power factor = sin ϕ—where ϕ is the phase angle between voltage and current.
(C) Power factor = cos ϕ—where ϕ is the phase angle between voltage and current.
(D) Power factor = V × I × tan ϕ—where ϕ is the phase angle between voltage and current.

45. Both inputs of the circuit in Fig. A-16 are at logic 0. The output:

(A) Is logic 1.
(B) Is logic 0.
(C) Cannot be determined from the information given.

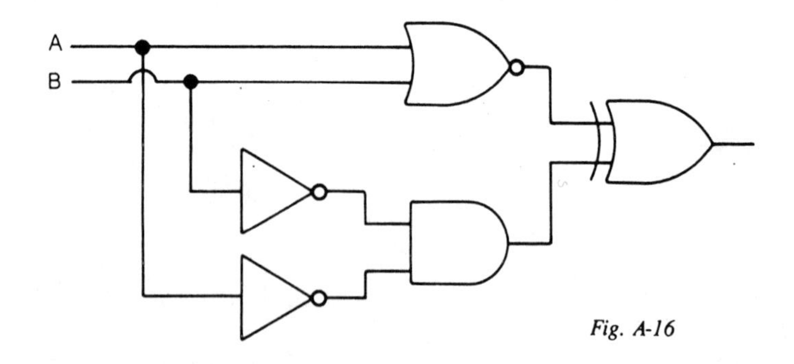

Fig. A-16

46. Which of the components represented in Fig. A-17 can be used as a power supply preregulator?

(A) The one marked A.
(B) The one marked B.
(C) Neither choice is correct.
(D) Both choices are correct.

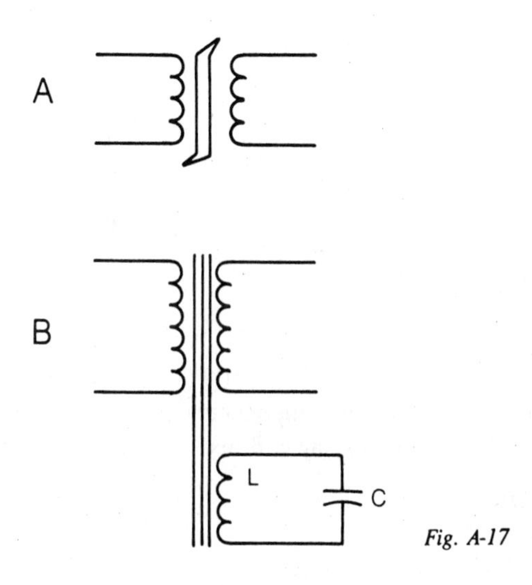

Fig. A-17

47. Which of the following is used for power amplification?

(A) VMOS.
(B) CMOS.
(C) JFET.
(D) Depletion MOSFET.

48. If the input signal is delivered to the gate and the output signal is at the source, the amplifier is:

(A) A follower.
(B) An inverter.
(C) A Class B amplifier with phase inversion.
(D) None of these choices is correct.

49. The op amp shown in Fig. A-18 is connected as a simple:

    (A) Differentiator.
    (B) Integrator.
    (C) Noninverting amplifier.
    (D) Inverting amplifier.

    (Important note: Make sure you know all of the op amp connections mentioned in this question.)

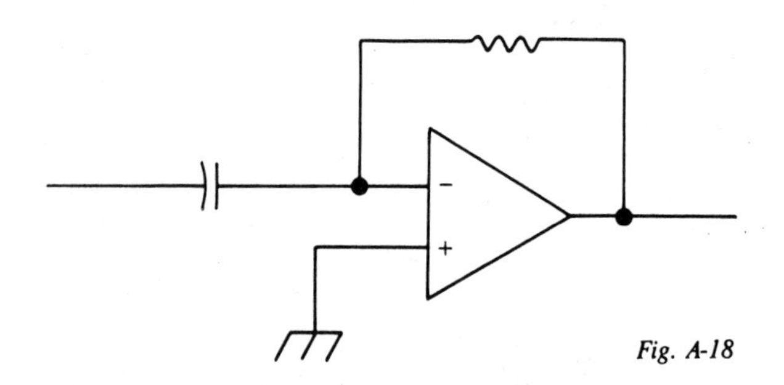

Fig. A-18

50. Which of the following components could be used to eliminate inductive kickback from a relay coil in a VMOS drain circuit?

    (A) VDR.
    (B) Diode.
    (C) Both choices are correct.
    (D) Neither choice is correct.

51. Form B contacts on a relay (or other switch) are:

    (A) Normally open.
    (B) Normally closed.

52. To interface CMOS logic with a 40-V display:

    (A) Use a directional coupler.
    (B) Use an optical coupler.
    (C) No special coupler is needed.
    (D) None of the choices is correct.

53. A shunt-wound motor:

    (A) Will operate on ac or dc.

(B) Will operate on ac only.
(C) Will operate on dc only.
(D) Has a higher starting torque than a series-wound motor.

54. An advantage of a relay over an SCR for controlling power is that:

(A) It is faster.
(B) It has a high input/output circuit isolation.

55. You would expect an electronic dc converter circuit to have:

(A) An oscillator.
(B) A duplexer.
(C) A synchronizer.
(D) A directional coupler.

56. In Fig. A-19 the driven wheel is A. Which of the following is true?

(A) A, B, and C turn at the same speed because of the idler wheels.
(B) A, B, and C turn in the same direction.
(C) B turns faster than C.
(D) C turns faster than A.

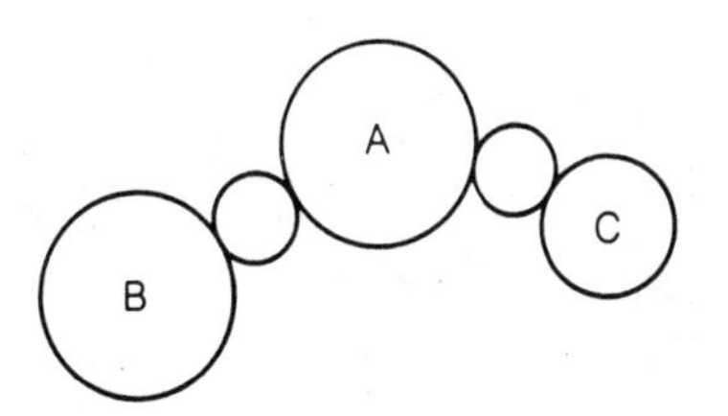

Fig. A-19

57. You would expect to find a snubber in:

(A) A converter.
(B) An inverter.
(C) A J-K flop flop.
(D) An SCR circuit.

58. Which of the following is not a thyristor?

    (A) SCR.
    (B) VARIAC.
    (C) DIAC.
    (D) UJT.

59. The components illustrated in Fig. A-20 are:

    (A) Depletion MOSFETS.
    (B) JFETS.
    (C) Enhancement MOSFETS.
    (D) CFETS.

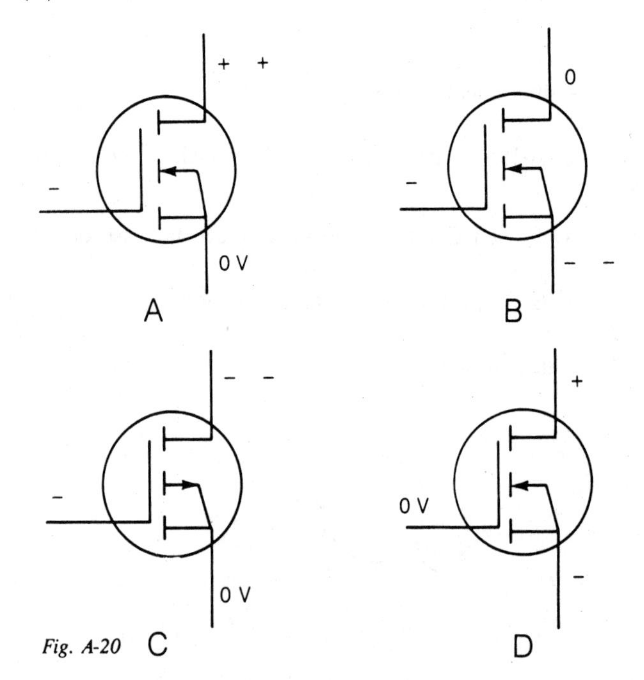

Fig. A-20

60. Refer again to Fig. A-20. Which of the components is improperly biased?

    (A) The one marked A.
    (B) The one marked B.
    (C) The one marked C.
    (D) The one marked D.

61. Refer again to Fig. A-20. Identify the N-channel device.

(A) The ones marked A, B, and D.
(B) The one marked C.
(C) All are N-channel devices.
(D) None are N-channel devices.

62. You are picking a component for a low-frequency clock signal, which of the following could be used for that clock signal?

(A) Toggled J-K flip flop.
(B) D flip flop.
(C) PUT.
(D) Triac.

63. A certain power supply has an output of 12 V when there is no load connected. With a full load connected the output is still 12 V. The power supply percent regulation is:

(A) 0%.
(B) 100%.

64. A program for a drill press is stored in machine language, which of the following could be used for storing it?

(A) Punched tape.
(B) Magnetic tape.
(C) Magnetic discs (called floppy discs).
(D) Hard disc.
(E) Any of these could be used.

65. The calibration of the scope is given with the waveform in Fig. A-21. Determine the RMS voltage and the frequency of the waveform.

(A) 7.1 V and 2 MHz.
(B) 7.1 V and 5 MHz.
(C) 17.4 V and 1.25 MHz.
(D) 10.6 V and 0.833 MHz.

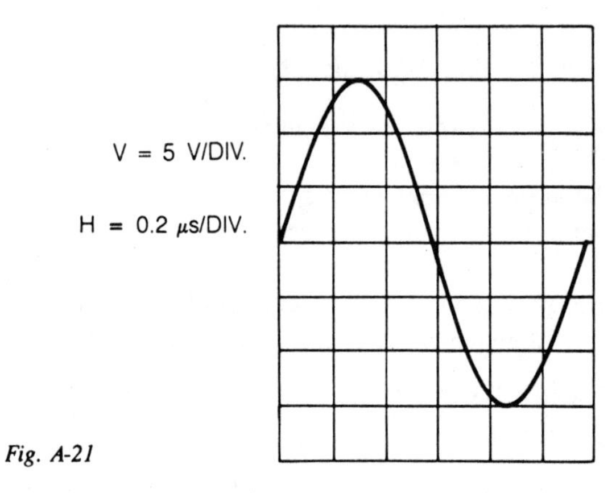

Fig. A-21

66. The circuit shown in Fig. A-22 is an example of:

    (A) A common collector amplifier.
    (B) A common base amplifier.
    (C) A common emitter amplifier.

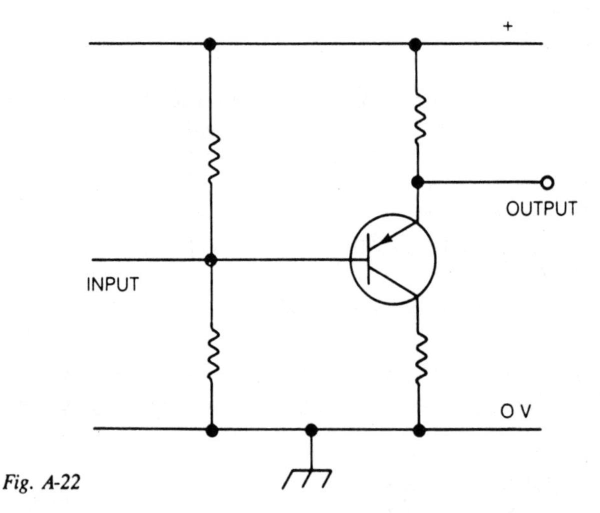

Fig. A-22

67. The operational amplifier in Fig. A-23 is connected for:

    (A) Maximum voltage gain.
    (B) Maximum current gain.
    (C) Common-mode operation.
    (D) Buffer operation.

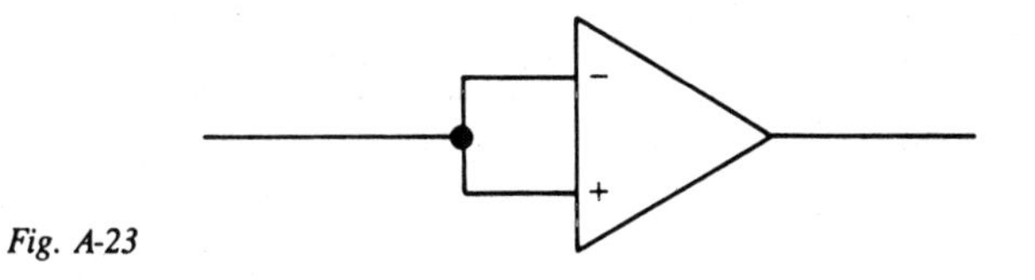

Fig. A-23

68. Which of the following is used to start large dc motors?

(A) A simple ON/OFF switch can be used.
(B) An SCR circuit (a thyratron circuit in older systems).
(C) A manual starter.
(D) None of the above choices is correct.

69. Each time a pulse is delivered to a certain motor its shaft turns 15°. It is:

(A) A synchronous motor.
(B) A motor with a shaded pole.
(C) A stepping motor.
(D) An incremental induction motor.

70. A Ward-Leonard control is used for adjusting:

(A) Oscillator frequency.
(B) Motor speed.
(C) Generator output power.
(D) Current in a varying load.

71. Which of the following is not a logic family?

(A) RCL.
(B) TTL.
(C) I^2L.
(D) CMOS.

72. The output of the op amp in Fig. A-24 should be:

(A) 0.2 V, peak to peak.
(B) 0.08 V, peak to peak.
(C) 0.5 V, peak to peak.
(D) None of the choices are correct.

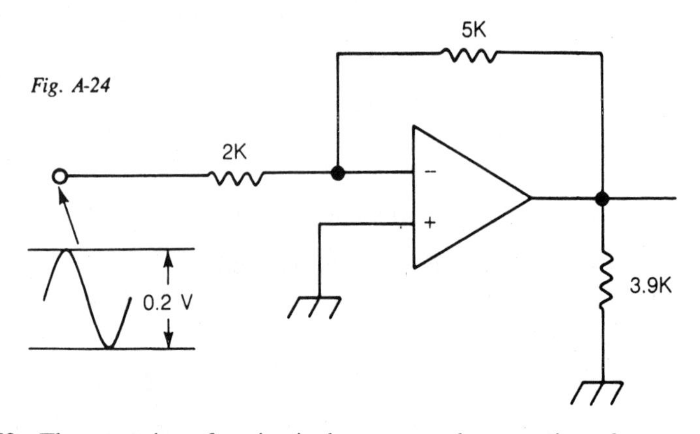

Fig. A-24

73. The operation of a triac is the same as the operation of:

(A) A thyratron.
(B) A diac.
(C) Back-to-back SCRs.
(D) Back-to-back neon lamps.

74. In a Class A bipolar PNP transistor amplifier, the dc base voltage is:

(A) Positive with respect to the collector.
(B) Negative with respect to the collector.
(C) The same voltage as the collector.

75. Which of the following can be turned ON and OFF with input signals?

(A) SCR.
(B) SCS.
(C) Both choices are correct.
(D) Neither choice is correct.

Answers to the Practice CET Test are located in the back of the book.

Index

ANSWERS TO PRACTICE CET TEST

Industrial Electronics Option

1. (C)
2. (D)
3. (D)
4. (A)
5. (C)
6. (D)
7. (B)
8. (D)
9. (D)
10. (B)
11. (C)
12. (A)
13. (B)
14. (C)
15. (C)
16. (C)
17. (A)
18. (C)
19. (C)
20. (B)
21. (D)
22. (D)
23. (D)
24. (C)
25. (B)
26. (B)
27. (B)
28. (D)
29. (A)
30. (D)
31. (C)
32. (C)
33. (A)
34. (A)
35. (A)
36. (B)
37. (D)
38. (C)
39. (B)
40. (D)
41. (B)
42. (C)
43. (C)
44. (C)
45. (B)
46. (D)
47. (A)
48. (A)
49. (A)
50. (C)
51. (B)
52. (B)
53. (C)
54. (B)
55. (A)
56. (D)
57. (D)
58. (B)
59. (C)
60. (A)
61. (A)
62. (C)
63. (A)
64. (E)
65. (D)
66. (A)
67. (C)
68. (C)
69. (C)
70. (B)
71. (A)
72. (C)
73. (C)
74. (A)
75. (B)